Land, Livestock and Life

A grazier's guide to finance

Land, Livestock and Life

A grazier's guide to finance

Allan Nation

GREEN PARK PRESS

A division of Mississippi Valley Publishing Corp.
Ridgeland, Mississippi, USA

First printing January 2007

ISBN 0-9721597-2-X

Cover Design by Steve Erickson, Ridgeland, Mississippi

Manufactured in the United States.
This book is printed on recycled paper.

Contents

Foreword

A paradigm has been described as the acknowledged rules for how to get ahead in a business or industry. In years past, in agriculture one of the fastest ways to wealth was to clear a forest, drain a swamp or as the saying went "make two blades of grass grow where formerly only one grew." Of course, the wealth came not so much from the increase in production but from the increase in the value of the land. This was because in an agricultural society, the nominal price of the land was set by its level of production.

Since World War II, North American farmers and ranchers have become extremely successful at increasing the gross output of the land and yet, today, virtually no one can pay for land despite its massive production. Why is this so?

What's going on is a fundamental shift in how a wealthy, post-industrial society looks at and values land. We are now a couple of generations removed from being an agricultural society even in our most remote areas. This means that the valuing rules of an agricultural society no longer apply.

For the first time in North American history, the average buyer of farm and ranch land is not expecting *any* production income from it. Its whole value is tied up in aesthetics and lifestyle. In this new economy, an occasional glimpse of wildlife is more valuable than any amount of livestock carrying capacity, and a pretty pasture is worth double that of cropped land.

The productivity and net return of the land is no longer

a major consideration in what today's land buyers are willing to pay for it. Unimproved rangeland fit only for goats and deer can actually sell for a higher nominal price per acre than irrigated crop land. The net result of this disconnect between nominal price and production value is that the more land you own today the lower your ranch's return on investment will become.

Unfortunately, most agriculturalists are still carrying in their heads the old paradigm of paying for land from its production. This has led to huge financial stress.

The rules of the game have changed but we haven't.

Trying to farm or ranch under yesterday's rules is akin to trying to butt down a four foot thick brick wall with your head. The only thing that will result is a severe headache. However, the situation is not hopeless.

Management guru, Peter Drucker, said that no industry ever really stops growing. When it hits an immovable obstacle, it stops for awhile, but eventually changes its paradigm to one that will work and starts growing in a different direction. North American agriculture is currently doing exactly this.

Today's farming game is more about obtaining control of agricultural land rather than deeded ownership. If you can understand exactly what today's land owners really value, you will find that obtaining the grazing use of land is actually easier and cheaper today than at any time in post-frontier North American history.

The livestock business is also undergoing a related paradigm shift. Today, the profit emphasis is on the *management* of grass and livestock more than the *ownership* of grass and livestock. Livestock ownership is both capital intensive and extremely risky. In contrast, the management of livestock requires only small amounts of capital, provides a monthly cash flow and involves virtually no market risk. The buying and selling of livestock management *services* is the new profit paradigm in today's ranching.

Of course, none of this shift in paradigms precludes you from owning land or livestock. It just urges you to do so on an incremental, piecemeal basis. In other words, *first* start a business. Then you can *choose* to invest your after-tax surplus in deeded farm and ranch real estate or you can choose not to. Either way, your livestock management enterprise can thrive.

Also, we must admit that we are affected by the same lifestyle issues as today's real estate investor. Like them, we too want a pleasant life in the country. We want a safe place to raise our children and clean air to breathe. However, the ultimate pleasant life in the country is not just a *life* on the farm but a *livelihood* from the farm as well.

The key point to remember is that this lifestyle is a *result* of having gotten the first two of these elements right. Far too many ranchers and their wives are working long hours off the farm to subsidize their country lifestyle because they have not gotten the financial structure of their farms and ranches correct.

Land, livestock and lifestyle —The b all important elements in a grass farm or ranch. Hopefully, this book will help you better understand how they fit together in today's modern economy. Read on!

Land

As an investment for after-tax dollars, farm and ranch land has been exceptional since September 11, 2001. Since that fateful day, many urban people have become increasingly disenchanted with urban living and see rural areas as safer places to live. Currently, 70 million urban Americans say they want to move to the country.

After dropping from dizzying heights similar to today's in the 1970s to a low of $599 an acre in 1987, the average price of farmland rose to $1900 an acre in 2005.

Farmland increased in value 15% in 2005 and 21% in 2004. In the same two years, corn prices were at record lows due to chronic overproduction. This shows the growing disconnect between land price and production profit. And, this is even worse in pastoral properties.

Some of the sharpest increases have been in the semi-arid to arid pastoral states where land productivity is often marginal. In 2005, Montana reported a 49% increase, Idaho a 40% increase and gains in excess of 40% in Nevada, Utah, Arizona and New Mexico. Nationally, pastureland increased from $820 to $1000 an acre in 2005.

Gene L. Dunbar of Bank of America told *The Wall Street Journal* that while farmland still has some small connection between price and productivity, ranch land doesn't.

"Farmland is bought on its ability to produce, whereas with recreational ranches all the logic goes out the window," he said. "If it has a habitat, scenery, wildlife or water, there is no logic to what it brings."

We now live in a world where scenery and wildlife add more value than agricultural production. This is definitely a different game than our fathers played.

In this section, we will take a look at how one plays and wins at this new game.

The basic rules of the game — Finance

Most people buying ranches today are doing it because they perceive a ranch lifestyle as fun. Once people are fully fed, they will pay more for entertainment than they will for food.

This is why trying to pay for inflated ranch land solely with more agricultural production is so frustrating. An acre of land with no production sells for pretty much the same price as an acre of land producing 500 pounds of beef gain.

Under this scenario is there any leverage in superior production skills? No. Will you be able to buy a larger ranch because you are a better rancher than most? No.

The reality is that if you believe you must own every acre your cattle graze you are doomed to a marginal income. Land is no longer a production tool. It is a toy and toys sell for more than tools because we value fun more than work.

Does this mean there will be no profit in livestock production in the future? No, it just means we will have to sit down and learn the new rules of ranching whereby production is adjunct to a higher and more valuable primary enterprise.

The reality is that raising livestock is not fun but work. A ranch is a 24/7, 365-day deal. Ownership of one cow means it's necessary for someone to be there in case she gets out.

The people who buy a ranch strictly for fun soon become disillusioned. If they get rid of the livestock altogether, the ranch grows up in brush and trees. Mowing just to maintain a landscape is work from the get-go. Remember, these people are into fun — not work.

They want the landscape but not the responsibility for maintaining it.

This is why people like Missouri grazier, Greg Judy, can often get the grazing rights to land for free. His clients want the fun of watching the cattle graze on their land but want to be able to jet to Hawaii for the weekend.

Now, if you can take the work out of land ownership *and* improve the look of the landscape with your grazing skills, these people will get really excited. If you take good care of other people's land, land will come to you. You won't have to go looking for it.

> *The mission statement for our ranch starts with, "Pursue the highest quality of ranch life possible today while building equity and cash flow for the future."*
>
> *It also includes certain principles I believe are critical such as, "...never forget this is a business, to do so guarantees failure."*
>
> *John D. Cox, Colorado Rancher*

While some people refuse to believe this will happen where they live, I have never seen it fail to work anywhere in the world. But, you've got to develop a reputation first.

How could you package and sell the entertainment and aesthetic value inherent in all rural properties to urban people? What is it that people are really buying when they buy a place in the country?

The number of people who can afford to own a sizable acreage in the country is limited but apparently the desire to own acreage is nearly universal.

How could you package this so that more people could buy the experience without buying the land?

Now, think about what you are doing here. You're thinking about other people's needs rather than your own.

This is called marketing.

It's not hard. It's not rocket science. But, it really pays well.This is the new ranching paradigm.

The rules of the game are called finance.

If you don't understand the basics of how the game is played and where the leverage is, you are going to work very hard and accumulate little.

The game your father successfully played was a very different one than you face. The game your children will have to play will be very different from the one you learn. The reality of today's fast-moving economy is that we will all have to learn a new game every three to five years.

As Confucius said, "A constant level of happiness requires frequent change."

One hundred years ago, ranches were good businesses because beef, mutton and wool were in short supply. People needed more production. Today, people don't need more of any commodity we produce.

What people need today is occasional relief from urban noise and pollution.

They need to feel the isolation of big empty spaces.

They need to feed their genetic longing for green landscapes and savannahs.

They need to reawaken their spirituality by rediscovering God in Nature.

When they eat, they don't want more. They want different.

They want food with an entertaining story behind it.

And, they increasingly want to know the people who feed them.

In a world suffering from an oversupply of things, what people want more of today is happiness. The ranchers who understand this and who devise ways to sell this most priceless commodity in all of its many forms will be the winners in the new ranching game.

Real estate, formerly known as farmland

It is estimated that the only rural land that is still affected by the profitability of farming and ranching is in the tier of states from the Texas panhandle northward to the Canadian border.

In all the rest of the country, rural real estate is now being priced by the urban economy.

In Texas, for example, the average land cost per cow in 2003 was $18,000. This means to earn a 5% return (the equivalent of a no-risk Treasury bond) a cow would have to earn $900 per year.

To pay the interest on a 7% mortgage, she would have to earn $1260 and that's just the interest.

Now, I thought this Texas figure was pretty extreme so I checked it against land prices in Mississippi where urban real estate is way under the national average.

Figuring the Southeastern average of four acres per cow, the land cost per cow in rural Rankin County, Mississippi, was $28,000 per cow!

In Northern Virginia, farmland sells for $30,000 an acre or $120,000 a cow.

In New England or California, these per cow prices easily double, triple or quadruple.

The average farmer has to be able to pay the average price for there not to be a bubble.

Even at record high cattle prices, the "average" rancher is making less that $100 a cow.

Therefore, one might conclude that the biggest real

estate bubble in America may not be in Miami Beach but in your own pasture.

I predict this bubble is likely to be much more longer-lived than the urban housing bubble.

The reason for this is that these rural land prices are being driven by the people who are wealthy enough to be unaffected by any dramatic shift in lender sentiment or level of interest rates.

A beautiful rural estate is a prized part of most wealthy people's personal portfolio.

These people want the same "quality of life" that we so prize and have long been willing to subsidize.

In 2005, ninety percent of American family farm income was from off-farm jobs, other businesses or Social Security.

The big difference is that the wealthy don't have to worry about how to pay the mortgage. They can and do pay cash for their toys.

Here's statistical proof.

The wealthiest one percent of Americans generate 33% of its taxable income but only holds 6% of its debt.

The very rich typically only use debt to build or buy assets that generate self-liquidating cash flow. They do not use it for lifestyle items including homes and rural estates.

This makes them largely immune to the interest-rate-driven economic cycle.

Their primary desire for acreage is not for production income but as a noise and social buffer from other people. They are, in effect, buying a rural "island."

This "island" could be 100 acres in Northern Virginia, 1000 acres in Northern California, 10,000 acres in Montana or 100,000 acres in Nevada.

Forbes Magazine estimates that the average rich guy is willing to spend as much on his recreational real estate as he is for his primary residence.

This means that if his primary residence goes up by

30% he is willing to pay that much more for, or buy that much larger of a farm or ranch.

A study of this "gentrification" phenomena found the wealthiest people tended to buy larger farms and ranches farther away from their primary source of income.

The millionaires have "gentleman farms" in Northern Virginia. The billionaires have huge ranches in the wilds of Nevada and helicopters to take them there.

If you live outside of the previously mentioned central tier of states, the odds that you will sell your farm or ranch to another working farmer or rancher is infinitesimally small.

A survey of working ranchers in Northern California by the University of California at Berkeley, found that while the majority of ranchers would like to sell to another rancher, next to none were interested in any kind of legal vehicle to restrict the sale of their land and thereby lower its price to the point where this would be feasible.

In effect, land inflation has divided farming and ranching into the same two-class system as seen with homeowners.

Farmers and ranchers who bought in during the 80s and early 90s are sitting pretty, however the next generation of production farmers or ranchers are effectively shut out.

Let me clarify that.

The next generation are effectively shut out as long as they try to farm or ranch under the same paradigm as the previous generation.

The old paradigm said you had to own all the land you farm or ranch.

The new paradigm says you want *control* of the land you farm or ranch, not necessarily ownership.

All we want is the grass. Not the ownership of the dirt that grows it.

Let's take a look at the new agricultural paradigm.

Consider that Northern Virginia farm at $30,000 an acre. At a three percent tax rate, the taxes on the estate will be $900 an acre each year.

An agricultural tax exemption that cuts this rate to a few dollars an acre is worth a small fortune to a landowner with a sizable acreage.

I do not know what the Virginia tax rate is and this is purely a hypothetical example, but the idea is widely applicable.

The point is that keeping up an appearance of farming to avoid taxes is a key element in these landowners' quality of life.

While these guys don't mind investing a few million in a farm or ranch, they hate an on-going cash drain.

Now, consider that the average wealthy guy will spend no more than 21 days a year on his farm or ranch. There is no way he has the time to farm or even keep the property from reverting to forest in the East.

> *According to* Forbes Magazine, *average farmland values have risen 1.1 percent a year above inflation since the Depression. This compares to 2.7 percent for the Dow Jones average.*
>
> *However, after peaking in 1981 they dropped by 35 percent by 1987 and have since just barely recovered, in nominal terms.*
>
> *In today's purchasing power, an acre of farmland is only worth two-thirds of what it was at its peak in 1981.*

And that's where you come in.

If you can solve this guy's two problems of keeping the property open and visibly in agriculture, he should be willing to pay you to do it.

Rich people do not buy islands in the Caribbean for the coconuts they can grow there. They don't buy islands in the countryside for their production potential either.

What they want is peace, quiet, and a pretty landscape with as little personal hassle and on-going expense as possible.

Land and livestock

Ranching is an incredibly complex "whole" made up of a lot of separate businesses that you must be equally good at to make a profit.

Here are just a few of those businesses:

You have a forage production business.

You have a genetics business.

You have a real estate business.

You have a livestock marketing business.

What we are seeing in studying the innovators among us is that the ones who are the most successful — the fastest — are the ones who tackle only one of these businesses at a time.

I believe this is the reason why some people can own land and make money leasing it to others but can't make money owning their own cattle.

It's why some people can own cattle and make money leasing land but can't make money owning it.

It's why some people can make money growing and selling hay to others but lose their shirts trying to make it for themselves.

It's why some people can sell gain for 32 cents a pound and make more money per head than the "average" of people who own the cattle.

And, it's why some people can pay 32 cents a pound and make more money than "average" people who grow calves for their own account.

The reason we are a service economy in the USA is that specialization increases the efficiency of both the buyer and the seller of the service. The most successful among us appear to have figured this out. As uncomfortable as it may be, perhaps we should put a pencil to it as well.

A great many North American ranches are incorporated as "land and livestock" companies.

Even though many of us typically have more money invested in real estate than in livestock few of us really understand the real estate business.

Prior to the great grain price inflation of the early 1970s, rural land prices were relatively closely tied to both the land's individual productivity and farm profitability in general. However in the inflation-wracked 1970s, farm and ranch land began to be touted as a good inflation-hedge for wealthy urbanites.

Grain prices came down in the late 1970s and farm and ranch land prices followed in the early 1980s. Many newcomers to real estate have ignored this big drop in farm real estate prices. Farm and ranch land prices are just now re-approaching the prices they brought in 1980. In some areas, they are still far below the peak reached in the early 1970s.

Since the mid-1980s, the price of farmland has closely tracked the inflation rate of the urban economy and has risen slowly but steadily. Over the last 20 years this increase in the value of farm and ranch land in the United States has been almost identical to the urban economy's annual inflation rate.

In other words, the after-tax dollars put into farm and ranch real estate since the 1981 price collapse have maintained their purchasing power tax-free. It has been a good investment since 1981.

However, the tie between land prices and its production that was severed in the early 1970s has never been rejoined. As a result, putting money into land ownership has been a terrible business decision as far as farm profitability is concerned.

Land ownership has been a great investment but a terrible business decision? Yes.

This dichotomy can be seen in the fact that a ranch with no livestock on it brings virtually the same price as a profitable working ranch. As a result, very few ranches are sold to new owners as intact businesses. Typically when ranches are liquidated, the cattle and productive assets are sold separate from the land.

This separation reflects that farm and ranch land today is priced by urban values rather than rural economics.

A survey by Texas A&M's Real Estate Center reported in the Noble Foundation's *News and Views* Newsletter that rural land buyers value a property's hunting and fishing potential as seven to eight times more important than its agricultural productivity and three times more important than its conversion potential to ranchettes or homesites.

> *Fifteen percent woody cover that is well distributed across the landscape can support substantial populations of bobwhite, turkey and white-tailed deer. Land with reasonably good quail or deer hunting can earn at least $2 per acre relatively easily through hunting lease income.*
>
> *If fifteen percent woody cover causes a whole property to earn $2 an acre, the woody cover may actually be earning $13.33 an acre. It is difficult to net (cash basis) $13.13 per acre from a cow-calf enterprise over the long term on native range.*
>
> *Mike Porter, The Noble Foundation*

San Angelo goat and sheep consultant, Bob Steger, said west Texas was going through a significant shift from production agriculture to becoming a recreation area.

He said "the last cheap land in America" was being bid to ridiculous levels by urban hunters, and full-time ranchers were cashing out in droves.

He said San Angelo's reputation as the sheep and goat capital of America may be in long-term jeopardy due to this shift in land use.

Dalton Maddox agrees. He grazes sheep and sells real

estate in Colorado City near San Angelo. He said the national slowing in real estate prices has not been felt in Texas yet.

In 2006 he said pasture land in his area was selling for $150 an acre more than irrigated crop land. Irrigated cotton ground was $625 an acre compared to $800 to $900 for pasture.

"There's only one buyer for irrigated crop land in our area. If you want to sell, you have to sell to him.

"If you've got pasture, you've got a lot more potential buyers."

He said the big difference between today's land buyers and those in the 1980s is that today's buyers don't expect any income from the land.

"The bankers love these guys because the ranches are all paid for with off-farm income and are unaffected by drought and cattle cycles."

Many of his current sales have been 1031 exchanges and others are all cash sales from ranchers who have sold their ranches further east and have to reinvest quickly to avoid paying taxes.

He said a lot of land east of San Angelo that sold for $40 to $50 an acre ten years ago was now selling for $350 an acre.

"They come in and say 'I have $800,000 to spend on land.' They don't really care what you can grow on it."

The Real Estate Center reported in 2006 you could take cropland valued at $400 to $850 an acre in North Central Texas and plant it to native grass pasture and more than double its appraised real estate value to $850 to $1600!

The report said that wildlife-friendly native grasses increased the value of a ranch by $50 to $200 an acre over improved varieties like bermudagrass. However, putting in pasture of any kind doubled the value of cropland in North Central Texas.

Unfortunately, most farmers and ranchers have not understood this huge shift in basic agricultural economics. That a view of deer and quail add more value to land than cows, corn or cotton is a very urban idea foreign to production-oriented farmers and ranchers.

In our land and livestock businesses, the livestock portion of ranching is still about capturing free solar energy in plants and converting it into a dense, high protein, human food. What has dramatically changed is the real estate side of the equation.

The Argentine rule of thumb is that the value of the livestock and the value of the land to support those livestock should be roughly equal for a ranch to be a good business. This means that land should track the value of livestock and rise and fall in concert with it. This tie between land and production is the heart of a truly rural economy.

In New Zealand, which is a rural economy, land is largely valued on the number of stock units it will carry. If you can double a property's stock carrying capacity you can double its appraised value.

This is certainly not the true in the USA and may never be the case again. However, the lack of correlation between income earning potential and cost does not mean that owning a ranch will not be a good investment.

The Catch-22 of land ownership is that while farm and ranch land ownership can be a good investment vehicle for after-tax dollars, it is normally a very poor vehicle for generating those after-tax dollars.

Dr. Arnold W. Oltman, an ag economist with North Carolina State, said that while deeded land ranches have been competitively profitable as urban investments they are seldom competitively feasible as rural businesses.

Business broker and author of *A Basic Guide for Valuing a Company*, Dr. Wilbur Yegge, estimates that the average annual return of land extensive businesses seldom exceeds three percent. Not so surprising then is the fact that three percent is the average return of deeded ranches in the USA.

In contrast, many ranches on leased land show double-digit returns equal to that of the best run public corporations.

Dave Pratt of Ranch Management Consultants in Fairfield, California, told me that in his most recent rancher benchmark survey ranchers who were leasing 75 percent or more of the land they used were five times more profitable than ranchers owning 75 percent or more of the land they used.

The lesson here is that if you want to make serious money in ranching, rent or lease.

Rents live in the real world and are constrained by rural enterprise profitability, and they go up and down in concert with the industry's overall profitability.

In true rural environments where all income comes solely from farming and ranching, land values are similarly constrained but such economies are now extremely rare in North America.

> *The unfair advantage wealthy people have is they never feel pressured to "make money." They already have all the income they need.*
>
> *The poor man can become the wealthy man if he behaves as if he is already wealthy and scorns all investments that do not provide income.*
>
> *Richard Russell*

Thanks to better transportation and computer technology, almost all rural areas have people with urban-generated after-tax cash to invest. In the urban economy, rural land is valued primarily for its inflation-proofing value and recreation.

As Utah ranch consultant, Gregg Simonds, said of the wealthy, "They've got money. They want happiness."

How much is too much to pay for happiness?

Such urban values have allowed land prices to rise much faster than agricultural rents in all but the most remote and forbidding locations in the USA.

Another major problem is that raw land has no depreciation tax subsidy as do buildings and improvements. Land is classed as a non-depreciable investment, and non-depreciable investments normally require that they be purchased with after-tax dollars. In contrast, rent and lease payments are tax deductible and so can be paid out of the much larger pool of pre-tax cash flow.

Robert Kiyosaki, author of *Rich Dad Poor Dad*, said the only reason you would ever want to own real estate is because banks will loan money to buy it and will loan money against it.

He describes real estate as a good place to "park" excess profits from a business. This protects the cash from inflation and it can be borrowed out of the property if needed.

Using debt to buy assets that throw off cash flow is just smart business, Kiyosaki said in his book *Prophecy*.

"First you build a business and then you buy real estate," he said.

Getting this critical sequence backward lies at the heart of most agricultural finance problems.

Kiyosaki said it is not only farmers and ranchers who don't understand the real estate business, urbanites don't either. He said most people see buying a home as a good investment whereas Kiyosaki said it is really one of your worst investments.

His definition of an investment is something that returns cash to you on a regular basis. A home is a good investment to your lender. It is a liability to you.

> *People don't want to hear anything that requires them to change their minds. They are looking for a short cut or secret passage to wealth.*
>
> *Ray Charon*

He said the way to tell if something is an investment or a liability was to "follow the money flow." If cash flows to you, it is an investment. If it flows from you, it is a liability.

This disconnect between the return and the value of an investment is not unusual in the urban economy.

For example, the coupon from an investment grade

bond would not pay the interest and principle of a bank loan to purchase the bond. Many urban investors hold part of their wealth in gold and diamonds, which pay no return at all. In the urban economy, farm and ranch land is viewed similarly.

If the money invested in the asset does nothing but hold its purchasing power over time, it is viewed as a good investment.

However, all such investments require a source of surplus after-tax cash flow to be feasible.

Similarly, stocks, bonds, diamonds, gold and the land can increase in value over time due to unforeseen changes in the overall inflation and subsequent interest rate. This can result in minimally taxed long-term capital gains. The reason these gains are taxed at such a low rate is that the circumstances that produced these profits cannot be foreseen at the time of purchase.

In other words, the risk is so high that it must be offset by a favorable taxing rate to get people to do it.

Rental income is also tax favored in the USA in that it is not subject to self-employment taxes as long as it is not your primary income source. However, farmers and ranchers are specifically excluded from paying themselves rent from a self-managed enterprise.

Therefore, tax-wise it is far better to be a landlord renting land to others than to be an owner of land in one's own enterprise.

Under USA tax law, the highest taxes are paid on income from labor. These high taxes are why it is almost impossible to become wealthy without owning a business and/or rent-generating tax-sheltered real estate assets.

In 1973 when the American cattle industry went totally insane and declared war on their customers, my Dad knew it was time to cut and run. He dumped the cattle ($1.00 a pound) and most of the ranch for a small fortune ($4000 an acre).

Today, this land is appraised at only one-fourth the value per acre of what my father was able to sell the bulk of the ranch for 30 years ago. If I trace today's appraisal price back to the ranch's original purchase in 1954, I find that it has returned around two percent per year to its original purchase price *before* factoring in inflation. This is not the way to get rich.

I am relating this to emphasize that "buy and hold" is probably no more appropriate a strategy in rural real estate than it is with cattle.

> *The prices of a few acres in the country are comparable to a postage stamp size building site in the suburbs. While we sit back and say you can't make money paying $2000 to $10,000 per acre for ag land...they are buying a building site and an alternative lifestyle that may seem priceless to them.*
>
> *John. D. Cox, Colorado rancher*

I read that real estate exists in three states - fully valued, falling in value, recovering in value. It does not go up (or down) forever, nor does it stay the same.

In other words, your farm or ranch real estate investment is every bit as cyclical as the cattle cycle but the cycle is longer at 25 to 30 years. As with the cattle cycle, what is most important is to have some rough idea of where you are within that cycle.

However, because most rural areas are now priced by urban values for the first time, the argument can also be made that "this time it is different" and that rural land values no longer have to have a tie to agricultural income and can just go up forever.

Keep in mind this is the same argument made about Internet stocks just a few years ago. The fact that rural land is priced so far above a "real" income floor price means that the risk in buying rural real estate at today's prices must be considered extremely high.

The old saw about if it is too high to buy, it is too high to hold, should be kept in mind.

The rules of the game:

1. If you want to make serious money in ranching, rent or lease.
2. Rent and lease payments are tax deductible and so can be paid out of the much larger pool of pre-tax cash flow.
3. Follow the money flow. If cash flows to you, it is an investment. If it flows from you, it is a liability.
4. Tax-wise, it is far better to be a landlord renting land to others than to be an owner of land in one's own enterprise.
5. It is more tax-efficient to borrow the capital gains out of a real estate investment than to sell it.
6. Today's urban economy real estate is a buy, rent, borrow business.
7. Understanding and keeping the roles of land and livestock separate and in their correct order is absolutely essential to managing a ranch profitably in today's urban based real estate economy.

Leased land and custom grazing

Today, almost no form of real estate will pay for itself from the rent it generates. This is true of rental houses, hotels, apartments, and it is certainly true of farm and ranch land.

If you read the real estate textbooks, this premium of price over income is a sure signal that you should be selling your real estate as it indicates a price "bubble." However, very few landowners have ever read a real estate textbook and most rural real estate is bought and held for emotional rather than economic reasons. Land today is too dear to buy, but is often extremely cheap to rent or lease.

This economic irrationality presents both a problem and an opportunity to young people trying to enter farming and ranching.

One young grass farmer who is seizing the opportunity of cheap grazing leases is Rucker, Missouri, grazier, Greg Judy. Over a three year period he was able to rent 1300 acres upon which he grazed 1100 stocker cattle on a custom basis.

Judy started out as most people do by buying some farm land just because he wanted it. He had a good town job so he never really put a pencil to anything. He just figured that cheap land would be easy to pay for. In the early 1990s, he put together around 200 acres of land from several small farms for between $300 and $400 an acre.

This land was what is termed in the South as "go-back land." In other words, it was former crop land that was trying to return to forest.

Judy said his acreage was severely neglected with large cedar trees and blackberry thickets. The topsoil had long ago gone down the Missouri River to points south.

To stock this land he bought 16 beef cows (at the peak of the cattle cycle) and found that he was totally out of grass by mid-summer.

For the first time he said it dawned on him that this farm was not only not going to pay for itself but that it was going to seriously put a drain on his off-farm, after-tax income.

> *Land Ownership:*
> *1. Lowers your overall return on investment.*
> *2. Lowers your cash flow due to lack of depreciation and deductibility.*
> *3. Is a one-time cost that normally appreciates.*
> *4. Can build equity without taxation.*
> *5. Cost is minimally tied to enterprise profitability.*

With a full-scale drought on, he happened to notice that a neighbor who had subdivided his land into paddocks still had plenty of lush, legume-rich grass. The neighbor told him he had learned how to do this at Jim Gerrish's Grazing School in Linneus, Missouri. (Gerrish has since moved to Idaho.)

Judy signed up for the school and slept in the back of his pickup for three days because he couldn't afford a hotel. He said the school was the best three days of his life and gave him new hope.

At Linneus he was also introduced to *The Stockman Grass Farmer* magazine. He got a subscription and found that its editor was screaming "sell mature cows" because the market was on the verge of collapse.

Judy screwed up his courage and in 1996 he pulled the trigger on the cows.

"It was the hardest thing I ever had to do, " he said. "I loved my cows, even though they were starving me to death."

Judy said that the total collapse of cow prices a few

weeks later helped him get over his heartache.

He took the money he got for the cows and bought stocker steers. Thanks to the profit from the steers he made his ranch land payment for the first time without dipping into the after tax income from his day job.

His elation was short-lived because that was the precise time his wife chose to take a hike. Now he had both divorce and land payments to meet and he figured his short ranching career was over.

"I was totally broke," he said.

Just when he was about to give up, a neighbor asked him if he would be interested in grazing some steers on a gain basis for him. Judy jumped at the chance. Using only the improved parts of his farm, he was able to put 200 lbs on the steers in 100 days.

Unfortunately, most of his farm was not improved and not of stocker quality. He said he bought an advert in the local paper looking for cows to graze. The advert didn't attract any cows but it brought him a herd of bucking broncs from a rodeo contractor. In retrospect, the horses were exactly what he should have been advertising for because they cleaned up the brush on the farm as good as any goat herd and made the land payment.

Custom Grazing:
1. Allows you to buy land and management on a per head basis.
2. All costs stop the day the animal is sold.
3. Provides a better cash flow than ownership.
4. Reputation business dependent upon grazier skills.

He said his paradigm shifted when he read an editorial in the *Stockman Grass Farmer* that said the purpose of farming and ranching was not to own land but to making a living from land.

"The whole article hit me like a brick in the forehead," he said.

Once he stopped thinking about owning land and started seeking to rent it, Judy said things just fell into place for him. Just three years after his change in mindset he was able to custom graze 1100 steers on 1300 acres. This acreage represents eight different farms, which are leased for prices that range from a low of free to a high of $20 an acre.

"I found that for no more than $400 a year I could control $60,000 worth of land. That is less than the interest on the land would have been without figuring any principle reduction," he said.

Ironically, once he stopped trying to own land he found that he could easily pay for his deeded farm. In just two years of custom grazing he paid off both his house and farm mortgage.

Now, about that free land. Judy said that the absentee landowner was so pleased with the way that Judy kept up his farm that he returned all of his previous rent money and gave him a lifetime grazing lease on the farm. That is for Judy's lifetime. Not the landowner's.

"It's been said that if you do a good job managing land, land will come to you and that is absolutely true. All you need is one really well managed farm that you can take prospective clients to and you've usually got yourself another farm."

Judy said that most elderly and absentee landowners are not really interested in the amount of rent they receive. They just want their land kept up with no continuing monetary outgo on their part.

"In this climate if you just leave land empty, it all goes back to brush and briars. If my cattle weren't there they would have to pay to bush-hog it."

One of his favorite tricks is to do a paddock shift while the prospective landlord is there.

"Seeing a large herd of cattle slowly walk through a gate with no hassle makes you look like a true guru to most people — particularly if they owned cattle at one time.

"'How long did it take you train them to do that?' one

of my potential landlords asked. When I tell them about three days, they are sold."

He said the ability to approach strangers and talk to them about leasing their land requires an honest, positive and optimistic attitude. He said the best way to develop an optimistic outlook was to take on small challenges and be successful with them.

"Once you have tackled a goal and are successful at it, it builds tremendous self-confidence. Once you learn the proper ingredients to be successful, just keep repeating the recipe.

"The beauty of custom grazing on leased land is that you can get scale quickly with no capital," he said. "I was able to climb out of my financial hole pretty quick.

> *Success comes from doing ordinary things exceptionally well.*
> *Warren Buffett*

"Also, you get paid each month rather than once a year as is typical with cattle ownership so the cash flow is better and the best part in these times of expensive calves - there is no market risk."

Judy said that custom grazing was similar to having a town job with regular pay and no market risk. He said that he doubted you would get rich doing it but you could make a very good living.

However as good as custom grazing is, he plans to use it as a stepping stone to cattle ownership. Judy plans to refinance his deeded farm at the bottom of the cattle cycle and use the money to buy his own herd so he can capture the capital gains of the rising market cycle.

He does not plan to buy more land. Here's why.

"I have no property taxes, no insurance. I have no risk of dropping land prices and land leases are 100 percent tax deductible whereas land payments aren't.

"Plus, acts of Nature, tornadoes, and floods will not break me. It's a mess to clean up but I'm not out any real financial loss."

Whereas, changing his real estate paradigm had a major role in turning his life around financially, equally important was learning to think years in the future and to use written goals.

"I have found that any goal is attainable as long as I stay focused on it. If you truly believe it will happen, by gosh it will."

The rules of the game:

1. The purpose of farming and ranching is not to own land but to make a living from land.
2. Custom grazing on leased land allows you to get scale quickly with no capital.
3. Custom grazing allows you to get paid each month for better cash flow than once-a-year payments with cattle ownership.

Needed: More good landlords

American ranchers' average age is 58 because most have built a successful off-farm business first. Today's average age is not radically older than he or she was 40 or 50 years ago, nor is it much older than the rest of the developed world's ranchers. Australia's average age is 68.

Many middle-age ranchers see their ranches as investments, not businesses. Having built their businesses, they now want to enjoy life. Most have neither the time nor the inclination to ever become good stockmen and are not interested in management-intensive anything.

While these people will seldom if ever become good ranchers, they could be good ranch landlords.

As a ranch landlord, they could have everything they bought the ranch for: The big country house for weekend escapes and to impress their city friends. The pastoral view. A barn for a horse or two. A view of cattle grazing in the fields, deer and elk at the edge of the woods. All this plus a steady monthly rent check from the grazier.

An entrepreneurial economy needs landlords. A young rancher can't start out buying land. He must rent.

Landlord status is not only preferable for part-time ranchers. What many working ranchers overlook is that there comes a time when it would be better for them to play landlord or banker as well.

New Zealanders practice sharemilking, a win-win situation for the aging owner in that it allows him to shift from earning money from labor to earning it from capital at a relatively young age. The goal in New Zealand is to be out of day-to-day dairying by age 50.

Because New Zealand beef and sheep ranchers typically sell — rather than will their ranches to their children — they also are out of the day-to-day much earlier in life than is typical in North America.

The ultimate — Being paid to graze

Pasture land rents typically move in inverse relationship to land prices. In other words, as land values rise, rents often go down.

Here's why.

If the underlying value of the property is rising, most real estate investors care less about rent because they are earning what are known as "portfolio" profits. Under USA tax law these portfolio profits are untaxed until realized.

As a result, most professional land investors prefer to "harvest" these profits by borrowing the equity out of the land rather than selling it. This produces no taxable income or sales commissions. Also, maintaining little equity in real estate deters bankers from foreclosure during the inevitable downturns.

As Dr. Willam J. Poorvu, the head of the Harvard real estate school said, "Real estate is not a buy and sell game. It is a buy, hold and borrow game."

He said that real estate should be thought of as a game rather than a business due to its frequent lapses of good business sense.

In a "hot" real estate market, a tenant who is willing to take care of, or improve the looks of a property will make the land owner far more money in untaxed appreciation than they could ever get in taxable higher rents.

The net result is that competent graziers with "pretty" properties that showcase their land improvement skills have been able to get pasture land for free, or more frequently, for just the taxes. But even free land is not the sweetest deal. Getting paid to graze is.

Goat graziers are showing the future of livestock landscaping by getting paid to graze. Eventually cattle will too.

Nebraska grazier, Chad Peterson, believes the highest and best use of ruminant animals today is to create attractive landscapes with fee-paid controlled grazing.

While this is considered a radical idea by most, Peterson, who is still in his early 30s, has been a pioneer in new pasture-based ideas his whole life.

His family was one of the first in the country to get into the buffalo business in the 1930s.

"We were grassfed before grassfed was cool," he said.

Young Peterson rode the buffalo boom in the 1990s to the top, cashed in, and switched to cattle when cattle were cheap.

He rode cattle prices up, cashed in, and switched to contract grazing.

He was the first in his area to get into meat goats, and now is pioneering fee-paid grazing for vegetation control and wildlife habitat management.

Did I mention that along the way he accumulated 9000 deeded acres in the Nebraska Sandhills?

A ranch improvement idea was to add a herd of meat goats to eat the broad leaf plants that the cattle and buffalo left behind. Two thousand goats were brought in from Texas and quickly made short work of the ranch's weeds and brush.

"The problem with goats is that they do too good of a job. They soon eat themselves out of their forage resource."

This led him to fee-paid grazing.

His 2800-head goat herd is now being grazed on a per head fee basis on a 4000-acre ranch owned by an irrigation and power district to remove weeds and brush on the Platte River Island in southern Nebraska.

"I'm known as 'goat guy' down there," he said.

Irrigation districts and hydro companies who use the

public's water are now having to provide waterfowl habitats as a mitigation fee.

"They have been trying to use fire and herbicides and have created a real environmental mess with both."

Peterson is convinced that landscaping with livestock will be the next big thing in grazing but will require some tweaking.

He has found there is an offsetting cost when you start using reproduction class livestock as tools.

"I have been very disappointed with the reproductive rates of my goats. The grazing fees are where the money is."

Farming to preserve the land

Phillip Hanes said he grew up sharing his father's distaste for farming but developed a deep love for the green mountainous landscapes of northern North Carolina and southern Virginia and the scenic New River than wound between them. He and his wife, Charlotte, had both spent their childhood summers there and had a great emotional attachment to the region.

"In the early 1980s, I had an epiphany," Phillip said.

"I suddenly realized that this pastoral landscape I so loved was created and maintained by farmers and their livestock.

"And, if the farmers weren't there, it would all go away and be converted into vacation home sites."

In 1983, he started buying pastoral farm land in southern Virginia and leasing it out to local farmers to graze free of charge.

"I just wanted to preserve open landscape and the sense of community.

"Unfortunately, the local people had no knowledge of rotational grazing and quickly destroyed the pasture with unmanaged continuous grazing.

"I realized then that we were going to have to get into farming ourselves to preserve our land."

In 1986, Phillip and Charlotte spent three weeks living with grassland farmers in New Zealand and studying their profitable methods of managing grassland.

"We realized that you can't push people,"Charlotte said.

"You have to show them."

In 1988, they bought another farm that came with a registered Angus herd. They found the registered business to be an unprofitable distraction and decided to go strictly commercial.

In 1992, they reoriented the 1200 acre ranch in a commercial direction to make it financially self-supporting. The ranch was reorganized as River Ridge Land and Cattle Company.

"We wanted to show our neighbors that grassland farming can be very profitable when it is done right," Charlotte said.

Mountainous Grayson County, Virginia, is the largest beef cattle county in Virginia and has the second highest mountain in Appalachia. The highest mountain is just across the state line in North Carolina.

The region has long been noted for its cool summers with daytime highs in the 70s and lush cool-season pastures. This cool weather and excellent pasture made southern Virginia and northern North Carolina the last major grassfed beef production region in the Eastern United States.

The ranch currently has around 650 Brangus and Hereford cross cattle. The only equipment the ranch owns is one tractor and a rotary mower. In grass surplus years, a neighbor cuts and bales hay for them on a custom basis to keep the pastures vegetative.

Despite their cold and snowy winters, they get by almost exclusively on stockpiled pastures.

Without any advertising, more and more absentee and retired landowners have called River Ridge to take over the management of their land. Four Grayson County farms have been leased and several others outside the county are currently under negotiation. River Ridge also has an active fee-based consulting business for landowners who prefer to manage their own land.

The latest initiative of the Hanes is to lead the region

back to its roots as a major grassfed beef production area.

"We notified our neighbors that we would no longer be participating in the graded cattle sale and would be going grassfed," Charlotte said.

"We told them they were welcome to come along."

Charlotte said their primary goal is still preserving the landscape and keeping farmers on the land.

"We just think grassfed beef is the best way to do that."

Real Estate — It doesn't always go up, a personal story

In 1954, my father paid $100 an acre for a small dairy farm in the Mississippi Delta south of Greenville, Mississippi.

This was the highest price per acre that had ever been paid for a farm in the Delta and earned my father the undying enmity of our neighbors.

These large acreage planters correctly foresaw that the severing of land cost from its underlying cash flow would destroy their then workable business model.

In 1954, three hundred acres of cotton could produce a lavish lifestyle on paid-for land, or with a slightly frugal lifestyle, one could afford land payments.

Of course, the reason Dad was willing to pay such an exorbitant price for a small farm was because it gave him the grazing rights to a thousand acres of Mississippi River levee and batture. He had had all the cotton farming he wanted growing up on a hard-scrabble North Alabama mountain farm.

He promptly leased out the 40 fertile acres of "ice cream" cotton land to a neighbor and went into the beef cattle business "where the cattle did all the work" as he put it.

By 1960, our little 200-cow ranch was producing an annual net income of $20,000. This is the equivalent of around $160,000 in today's dollars and produced an equivalent lifestyle for our family.

I want to tell you what happened to that 40-acre cotton field and why you may want to think before you start selling off small pieces of your ranch.

This 40 acres of plowed down dairy pasture produced a

good two and half bales of cotton year-after-year of which we received half as payment.

My Dad had grown up in Appalachia where he had struggled for years to make a half bale of cotton per acre. He saw getting a bale and a quarter for "nothing" was the only way to farm cotton and we stuck solely to beef cattle.

In 1973, the great agricultural boom hit and overnight seemingly everyone in the Mississippi Delta was a paper millionaire and began to live accordingly. In this boomtown atmosphere, a leading residential developer in Greenville approached Dad with the idea of buying that 40-acre cotton field and turning it into an upscale residential community.

Few planters' wives wanted to live on a remote plantation anymore. They wanted a large "town house" in socially active Greenville and many planters commuted from the city to their plantations each day.

After a considerable period of negotiation, Dad walked away with several hundred thousand dollars in return for a deed full of covenants that he thought would ensure an attractive looking real estate development and would protect the remainder of his property.

And he died suddenly and unexpectedly in 1974, thinking that was so.

The remaining property was a large brick house and a sizable pecan grove with pasture beneath it that we used as a bull and horse pasture and for shipping and receiving cattle.

Well, you students of history know that by 1975 the agricultural boom had turned bust and with it went the whole economy of the Mississippi Delta. The developer we sold the property to, and the bank that financed it, got sucked down in the subsequent massive "de-leveraging" that occurred over the next ten years. The net result was that the "upscale" development got downsized to attractive, but small, beginner-size homes built at a very high density.

Now, fast forward 30 years to when I became "executor" of my step-mother's estate, which by then had been reduced to the house and remainder of Dad's original farm. By this time the adjacent subdivision had deteriorated into a slum and a home to druggies of all kinds. I had the property appraised and was not surprised at its low appraisal.

However, my deceased stepsister's kids had seen an appraisal made 10 years before that was twice as high. "Real estate has to go up. It can't fall in half," they insisted and refused to sell for less than the ten-year-old appraisal.

It is really hard to convince kids who live in Dallas that yes, real estate can fall in half but it does.

The property finally sold in July of 2004 for virtually what the most recent appraisal had been. Or, half of its 1993 appraisal and a quarter of what it would have sold for in 1973 if Dad had sold the whole ranch. And, I felt extremely lucky to get what we got.

There are a couple of points I want to make here.

One is that real estate does not have to go up over time. It all depends upon the point in time from which you start measuring it.

Two, real estate values are the highest in anticipation of change rather than after the change has occurred. Your first offer is often the best offer you will ever get.

High density, suburban developments, such as what our cotton field became, do have a tendency to become slums over time and they will take down the value of any undeveloped adjacent property.

The value of your property is highly dependent upon what your neighbor does with his. Let him put in a large confinement hog operation and see what happens to the value of your land.

You may sell a piece of your ranch to a nice person who may then resell it to an ogre who makes your life miserable.

In conclusion, you can sell a small piece of property for what appears to be an excellent price but that piecemeal sale could actually cost you far more than you receive by lowering the value of the rest of your property.

Sell it and keep it

The Maytag Ranch is a 2953-acre ranch at the east end of the Royal Gorge between Westcliffe and Salida, Colorado. Billed as "An Agricultural Preservation Community," Russ and Jeanie Maytag are planning to preserve a working cattle ranch by selling the "lifestyle" portion of their ranch to others.

Largely sub-irrigated with a majestic view of the 13,000 foot Wolston Mountain at its rear, the ranch is both highly productive (1000 yearlings in the green season) and extremely scenic. Organic grassfed beef is the ranch's primary production emphasis.

Maytag said he loved living on the ranch but wanted to cash in on Colorado's popularity as a prime second home and retirement site.

"I wanted to pull my built up equity out of the ranch but still have a nice place to live," he said.

He said he had been following New Mexico rancher, Jim Winder, and his emphasis on conservation minded "soft" ranch real estate development for several years.

"I really liked the idea of being able to sell the ranch but, in essence, still keep it."

The ranch has been divided into 27 "ranches" sized between 80 and 100 acres. These are priced just north of a million dollars each.

The land is sold as "owned and shared," which means that the land is not fenced into individual parcels but is used in common. This means that while the cattle are primarily confined on the unsettled, sub-irrigated land, they can graze on any land owner's front yard.

Unlike some developments where very small lots are sold and the remainder of the land is put into a no-development conservation easement, Maytag said he felt the "owned and shared" concept was easier to explain to potential buyers than a conservation easement. However, only 120 acres of the 2900 acres will be impacted by any development.

The ranch development includes a horse barn and arena, an old time swimming hole, horse trails for riding, a lodge, shop building, two houses for the resident cowboys and two guest cabins for overnight guests.

All ranch community structures will be built from rough cut, naturally weathered lumber. Electric, phone and fiber optic lines have been placed underground.

A great deal of effort has gone into preserving large open vistas to give the feeling of living on a large acreage ranch. The house sites have been carefully chosen so that none can be seen from one another and are sited to be unintrusive on the landscape as possible.

"We let the land dictate the house sites. What we want to preserve most of all is the open landscape. There will be 25 families living here but their houses will be largely invisible."

Homes are limited in size to no more than 8000 square feet and will be connected to the main gate and each other by 10 miles of hard surface roads. Minimalist, portable one-wire electric fences are used for the interior fencing to be as unobtrusive as possible.

While there have been other ranch developments that emphasize maintaining the natural grassland landscape aesthetic, most of these separate the ownership of the livestock from the land.

Typically, the conservation land grazing rights are leased to a private rancher. However, Maytag said he envisioned "a community of ranchers" rather than a community of ranch houses and wanted the land owners to own the cattle as well as the land.

"I wanted them to have a feeling of ownership in what we were doing in the production end. If my cow poops in my driveway it's not nearly as offensive as if your cow poops in my driveway."

He hoped this would head off some of the grazier - land-owner conflicts other developments have had. Also, he thought the ranch owners would like to be able to describe the cattle on the ranch as "my" cattle to their urban friends.

Under the Maytag plan, the cattle will be managed by two full time cowboys for grassfed beef production and each ranch owner will receive their own personal supply of beef from the herd each year. Cattle surplus to the community's own beef consumption will be purchased by Maytag at market rates and moved to his second ranch near Pueblo. He estimates that this will provide $40,000 a year in income for the ranchers' association for additional improvements and upkeep.

In addition to the supply of grassfed beef, the ranch owners will receive fresh vegetables from the ranch garden, free-range eggs and milk and dairy products from the ranch's grassfed Jersey cow. (Green season only.)

This decision to include the land owners in the production side had a very serendipitous side effect. It allowed the ranch to borrow money for its development from the Farm Credit System rather than a commercial bank.

"The Farm Credit System is very enthusiastic about what we are doing here. They had wanted a way into financing real estate development and we gave it to them."

He said that a lender is an absolute necessity as there are sizable up-front costs in such a development.

The Maytags plan to continue to live on the ranch after it has been developed. He said this should be comforting to potential buyers.

"When the developer plans on living there himself, you know there won't be any corners cut."

Flint Hills ranch practices what it preaches

Kansas rancher, Pete Ferrell, describes his ranch's financial and management structure whereby land ownership and cattle are run as separate entities as just "practicing what I preach."

A long-time facilitator in the Executive Link program of California-based Ranch Management Consultants, Ferrell has separated the ownership of the land from the livestock business just as the Executive Link program recommends.

"Ferrell Ranch, the land owner, is one entity.

"4L Grazing, LLC, a management company, is another separate entity," he explained.

"4L Grazing rents the grass from Ferrell Ranch at market value for Bluestem pasture. I describe it as buying grass wholesale, adding a service and selling it retail."

He said this separation has several advantages.

"One, it is a clear statement to my children as to how they can best perpetuate this ranch.

"They can be active in management and work for 4L Grazing or they can be passive landowners and just collect rent.

"Two, it facilitates an open book style of management with my ranch manager, Jamie Nelson. He knows how, where and if we make any money.

"Three, it keeps us competitive in obtaining other ranch leases because we are used to paying it.

"Four, I believe that every asset you own should give you at least a 10% return.

"You should always charge yourself what the asset could earn as a bona fide investment.

"Five, it creates a legal firewall and helps keep your accounting clean and simple."

He said the long-range vision he has for Ferrell Ranch is to explore all avenues of sustainable economic development.

"Our motto is that we succeed because we recognize that ecology and economics are forever connected."

Some of the enterprises on the 7000 acre ranch are wind-generated electricity, custom grazing, hunting leases, agritourism, and, perhaps in the future, carbon sequestration and bottled spring water.

"I admit that I got totally burned out after 30 years in operations. Ranch management is really hard work and lots of it.

"My primary job now is to sit and think up new ways to sell stuff we've already got — like the wind that is blowing 250 feet off the ground.

"I have very little to do with the actual day-to-day cattle management any more."

Where is your money?

Capital in the ranching business is tied up in one of two things: Things we mean to keep and things we mean to sell.

The things we mean to sell are those that are used to generate profit and cash flow.

The things we mean to keep, on the other hand, fall into one of two categories — non-productive assets that make no contribution to profit such as houses, vehicles, boats and tennis courts; or productive assets such as land, orchards and breeding stock herds that indirectly contribute to profit.

By indirect, I mean they never are sold and do not directly contribute to profit.

When we buy something we mean to keep, money flows out unless someone else is paying us for the use of that something. If they are, it becomes an asset.

If no one pays for its use, it is a liability because owning it means that money must flow out in order to maintain the item. Toys are never cost-free.

If it is a true asset in the real sense of the word, money will flow into the busines thus increasing profit and net worth.

Some 30 years ago, I suggested to a farmer he would be better off selling the farm he had inherited and leasing it back. This would free up the capital he had tied up in his land so that he could use it to buy more things to sell and thus increase his net worth or equity.

To the best of my knowledge, he still farms the farm but he doesn't own it. What a happy position to be in.

Stan Parsons
Founder of Ranching for Profit

Livestock

During my 30 years as editor of *The Stockman Grass Farmer*, I've been privileged to meet people who were willing to think "outside of the box." Two men particularly stand out as teaching me the most about making money with livestock. One is Mississippi stocker grazier, Gordon Hazard. The other is Texas-based animal behaviorist and cattle marketing guru, Bud Williams.

I met Gordon during the bottom of the cattle cycle in the late 1970s. Almost everyone I interviewed that year was singing a song of woe, but Gordon declared that he was making lots of money. He told me that profitability was "a state of mind," that people who didn't constantly seek profitability would never find it. He said that in the commodity business it was always essential that one be "tight," but equally important, one could never be "cheap."

While Gordon Hazard is an extremely intelligent and thoughtful man, Bud Williams is a certifiable genius. Late in life, Bud moved to Alberta, Canada, where a young cattle feeder/grazier gave him a free-hand in marketing the cattle. From this experience he perfected his techniques.

Personality-wise Gordon and Bud were as different as night and day. Gordon is as down to earth as Bud is mercurial. However, both men were excellent students of the livestock market and share the belief that the only way to *manage* market risk was in real time.

Both men argued that what you paid for an animal six years ago, six months ago or six minutes ago was irrelevant because that price could not be changed. They emphasized you can only *manage* things that you *can* change.

In this section I hope to show that if you combine the marketing lessons of these two men with a study and reverence for how Nature works, you can make a profit ten years in ten, no matter what segment of the beef industry you are in.

Cycles, and how to ride them

Livestock profits are created in two ways:

One is by producing beef for less than you can sell it for and the other is by capturing inventory price increases.

The cattle cycle deals with the latter and only minimally with the former.

While the cattle industry is normally thought of as having three divisions — cow-calf, stocker-growers, and finishers — it actually has four. This fourth one is the harvest and processing segment.

Each of these segments has its own separate profit cycle. The cattle cycle we talk of the most is that of the cow-calf segment as it is the longest with an average of around ten years.

In 2006, the cow-calf segment was at its peak profitability, the stocker and finishing segment were struggling, and the processing segment was losing money by the bushel basket.

Now, where would you invest your money?

If you are like most people, you put it into increasing cow-calf production because it is currently the most profitable.

Most of us see the existence of high profits as the proof of the value of an investment, yet any student of economic history will find this is always the absolutely worst time to invest.

Why do we always forget this when prices are at record highs?

I found the answer in a book called *Bull!* by Maggie Mahar. This book is the history of the 1982 - 1999 stock market boom but is also a pretty good analysis of human economic behavior. "The human imagination tends to err in the direction of desire," she said.

The ultimate American dream is wealth without working at all. Profit from production always requires skill and tenacity and is therefore not nearly as attractive as catching a rising tide that is effortlessly lifting all boats.

Capital gains are also the lowest taxed profits and this increases their allure. Seeing other people achieving effortless wealth will always attract a large crowd willing to mimic the pioneers.

"People are social creatures, they are influenced by the spirit of their times," she said.

We invest with little thought or research about the underlying value or history of the investment because everyone else is doing it and we don't want to be left behind.

Mahar quoted Bill Seidman of CNBC, "You can never know what the American public is going to do, but you do know that they will do it all at once."

This subsequent "piling on" results in **all** investments eventually becoming over-valued and collapsed in price. The obvious strategy would be to always go against the crowd.

Forbes Magazine said they have not found a single instance where this didn't work. However, this is not easy.

What is so unnerving is that overvalued investments can remain hopelessly overvalued for quite some time.

"Market can stay irrational longer than you can stay solvent," said economist John Maynard Keynes about a contrarian strategy.

Once everyone who wants in is "in," the price increase stalls and then heads downward searching for its true intrinsic value.

Typically, this downward plunge erases the previous five years worth of accumulated capital gain profit.

The market will not turn around and start upward until, as Mahar puts it, "Every bull's heart is broken."

Of course, this is the time to buy but very few will do it.

According to Mahar, it takes about seven years for most people to psychologically get over a good economic spanking.

Also, if you have invested late in the cycle you will not have any surplus cash to invest at the bottom of the cycle.

As a result, most of us miss out on the lengthy upward side of the cycle and typically only get in again after the market is again becoming overvalued. In other words, they make the same mistake again!

> *It is what you learn after you think you know it all that is the most valuable.*
>
> *Clotaire Rapaille*
> *The Culture Code*

They rationalize away their past spanking by believing that there were clear signals of the past price collapse that they just missed. This time, they think, it will be different because they will watch more closely.

This seven year psychological lag explains the ten year nature of most economic cycles including the cattle cycle.

In other words, the ten year cattle cycle has very little to do with cows and everything to do with human psychology.

A key point that Mahar made is that a "buy and hold" strategy will not create nearly the capital gains that a "buy cheap and sell high" strategy will.

However, what most of us ignore is that this latter strategy requires that you have an alternative to re-invest these harvested capital gains into. The key point is, you don't put capital gains profits back into whatever just made them.

One of the best alternative investments is cash —T-bills, CDs, savings accounts.

Now, put this in the context of cattle.

Profits are highest when cattle prices are high. If you and your neighbors reinvest all of these high profits in more cows, this inevitably pushes cows higher than their intrinsic value. These super-high prices then attract new money to the industry, which bids up the price of cows even more.

Once cows bring more than finished slaughter heifers, almost all heifers will be routed to becoming cows.

This routing of heifers away from the abattoir creates supply shortages for them, and their profits collapse as they bid up the price of the remaining slaughter cattle to try to maintain capacity utilization.

This same heifer disappearance phenomenon drives up the price of feeder cattle as feedlots bid higher to maintain enough capacity to cover their overheads.

These investor-caused price increases are often misinterpreted as an increase in consumer demand. This misinterpretation is often deliberate.

Keep in mind, that in an over-valued market it is in no one's best economic interest to point this fact out.

So what can you do?

On paper, a successful strategy for the rising phase of the market would be to invest the profits of one phase into whatever you bought to make that profit.

In other words, finishers should reinvest their profits in stockers and stocker operators should buy cows rather than more stockers.

This reflects the truism that the majority of profits wind up being bid into the raw materials that made them.

> *You have to be small, smart and special or big, smart and fast.*
>
> *John Ellis*

Remember, that in the falling phase the losses move similarly but in reverse from sector to sector starting with finished cattle.

In other words, losses in the finishing segment get bid into lower heavy feeder cattle, which get bid into lower calf prices, which get bid into lower cow and seedstock prices.

This price unwinding of the whole beef segment normally takes about two years and this segment lag is what is the most confusing.

Early in the downward price phase cattle feeders can be

losing their shirts and calf prices still remain at record prices for a considerable time.

Cattle feeders typically will not stop paying high prices for feeder cattle until they have spent all of the profits they made during their profit years of the cycle.

For awhile, it appears that cattle feeders are immune to the pain of their losses and this lulls feeder cattle producers into a sense of complacency when they should be scurrying to get their economic house in order.

The point to remember is that your profits tomorrow are dependent upon the profits of the guy in front of you today. If he is losing his shirt, you eventually will too. So, pay attention to his bottom line as much as yours.

In a falling cattle market, the person with the lightest animal loses the least.

Bud Williams can show you that selling heavier animals and buying back lighter ones is a good way to ride a falling market down. However, this frequent selling strategy flies in the face of human nature.

Most of us freeze when the market drops and we ride all of our inventory price gains all the way down. This freeze reaction is so strong that Wall Street trading firms' computers are rigged so that they automatically sell a stock if it falls to a certain amount.

They know they can't trust even seasoned professionals to sell in a falling market.

I have personally learned that I have to harvest my capital gains profits when things are going well and I am not under duress. However, this strategy brings us into conflict with another great human desire.

We all want to sell one second before the market collapses. Not one minute and certainly not an hour before, but one second. Such precise market timing is impossible.

"The majority of investors agonize over the prospect of getting out too early and missing out on the profits that would have made them rich," Mahar said.

Warren Buffett, for example, completely liquidated his stock holdings in 1969 because he felt the market was grossly overvalued.

The market did not crash until January of 1973, a full four years **after** he sold out.

In other words, he sat on his wallet for four years. That is discipline!

He then stepped in and bought the then grossly undervalued stocks with abandon.

He said that the only way to have the cash to do this is to sell whenever prices become grossly over-valued.

That's the hard part. You actually have to sell to make capital gains real.

How do you know when to do this? Listen to your gut.

Bud Williams said, "If you feel cattle are too high for you to buy, they are too high for you to own.

"Everyday you don't sell your cattle, you buy them. There is no way to own cattle and just sit on the sidelines and not play."

The rules of the game:

1. Profit from production always requires skill and tenacity.
2. A "buy and hold" strategy will not create nearly the capital gains that a "buy cheap and sell high" strategy will.
3. A successful strategy for the rising phase of the market would be to invest the profits of one phase into whatever you bought to make that profit.

Lag in the cycles

In the spring of 2006, I spent an evening with a dozen or so Nebraska alumni of Bud Williams' Marketing Schools.

The general agreement at the Nebraska gathering was that the rumbling sound in the distance was not the bass drum of a brass band signaling a reprise of the Atkins Diet but the thunder of an approaching price storm.

While Bud taught that a falling market should be no more fearful than a rising market, a lot of us weren't so sure.

We pulled out our notes and reviewed Bud's advice.

Bud said the number one **wrong** thing you can do in a falling price market is to freeze and do nothing.

The faster you sell your inventory and replace it with a lighter weight category the smaller will be your drop in price and net worth.

This makes sense.

The heavier your animal the more at risk you are from a market drop because there are more pounds to decline in price.

A 400 pound calf that declines $10 cwt costs you $40.

An 800 pound feeder that declines $10 cwt costs you $80 or twice as much.

Bud said that in the late stage of the price cycle you need to really be watching what is happening at the abattoir as the slaughter cattle price is the **only** real price. And it is the only point that **real** money enters the livestock business.

You need to watch which direction the harvest price is going because it is the engine that pulls the rest of the train.

A falling harvest price eventually pulls all the rest of the industry down with it.

Note, I said eventually.

There can often be quite a lag between when the cattle feeder gets whacked across the nose and the cow-calf producer feels the pain.

Bud said all he needs to know for futures are the current cash cattle prices. He ignores futures prices because the quoted price is the price that no one thinks it is going to be because it is the average of people who think the market is going up and the other half who think it is going down.

> *If you are doing something right, most people will disagree with you. You can't be successful and popular.*
>
> *Bud Williams*

Cash prices are much more based in reality and exist as a way of communicating needs and wants between the various segments of the industry.

A rising price says send me more.

A falling price says send me less.

Again, what makes this so difficult in cattle production is the lag in prices between the different segments.

Generally, the cattle business is divided into three major segments. Cow-calf, stocker graziers and feedlot finishers.

Both profits and losses move through the segments like a slow moving wave.

A profitable situation in the segment you sell to forecasts a rise in the price of your class of animal eventually.

However, these three segments tend to be rather self-referential rather than holistic. In other words, losses in another segment are not seen as affecting them and little attention is paid to them.

Again, this is reinforced by the time lag between one segment's profit or pain and it reaching the next segment.

Let's look at the May 19, 2006 market average as reported by the USDA for an example of this lag.

Number 1 steer calves weighing 450 pounds were bringing $138.20 per hundred pounds of weight or $1.38 cents per pound of liveweight or $621.90 per head.

Similar calves weighing 550 pounds were bringing $126.50 cwt or $697.34.

Calves weighing 650 pounds brought $117.64 or $764.98 per head.

Yearlings weighing 750 pounds were priced at $104.63 or $781.72.

So, how much was the market willing to pay you to put on 100 pounds of gain that week?

For the hundred pounds between 450 and 550 it would pay $69.44.

For the hundred pounds between 550 and 650 it offered $67.64.

For the next hundred pounds between 650 and 750 it only offered to pay you $17.00.

That's right seventeen dollars for 100 pounds of beef gain!

While this may sound extreme, in mid-April when I figured this, the market was only paying eight dollars for this same 650 to 750 weight gain.

In this same May week in 2005 the market was paying $54 for this same hundred pounds of gain.

What does that low value of gain between calves and feeders tell you?

It doesn't sound like the feedlots wanted many feeder cattle on May 20 does it? Here's why.

On the same day, Choice slaughter cattle were bringing $79.37 cwt or $952.44 for a 1200 pound steer.

This steer when sold and replaced with the 750 pound steer at $781.72 produced a whopping 37 cents per pound of gain or roughly a $100 head loss on the swap.

In other words, a little figuring would have shown you that 750 pound feeders have to fall even farther in price before cattle feeders can break even and the market is clearly warning you about this.

Now, here again is where the lag comes in.

Cattle feeders typically have to lose money for two turns of cattle before they feel the whacking the market is giving them.

As a result, it appears to feeder cattle sellers that feeders don't really care about profits and tend to ignore their pain.

> *Entrepreneurs believe that profit is what matters most in a new business, but it isn't. Cash flow is what matters most.*
>
> *Peter Drucker*

In May, spring purchase stocker guys were ignoring this warning and buying calves like it was still 2003, because they had made good money **last** year.

Bud Williams and Gordon Hazard both use replacement inventory accounting to avoid making these blind leaps into red ink.

With replacement inventory accounting (known as last-out first-in at the IRS) you figure your price margin in real or present time.

The futures market is really designed for occasional players who are trying to maximize the profits on one set of cattle rather than an on-going business.

Last-out first-in accounting is recommended for all businesses with highly volatile inventory prices but many accountants have forgotten about it because they see themselves primarily as historians. History makes an interesting read but is of little use in managing a chaotic business in real time.

An analysis by *Cattle-Fax* found that in the vast majority of times when the initial sell-buy "swap" was negative at purchase the future buy-sell margin also turned out to be negative.

The big problem with buy-sell accounting is that in a rising market it can quickly put you in a bad cash flow situation.

With buy-sell accounting you can be reporting taxable profits on inventory bought months ago while your replacement inventory has dramatically increased in price.

This is why people using buy-sell accounting typically cannot maintain a constant sized inventory in a rapidly rising market from their own cash flow.

With sell-buy accounting, if you decide to buy another set of cattle after you sell a set, you know your profit or loss from the get-go. You don't have to wait six months to know whether or not you made a profit.

(The IRS allows you to change your accounting once. You cannot switch back and forth so as to minimize taxes.)

The problem with sell-buy is what do you do when there is no profitable trade?

This happens more and more as the profit cycle ages and is usually a sign that all of the profits have been bid to the cow-calf producer.

> *If it is too high to buy, it is too high to own. You must absolutely train yourself to believe this to be successful in commodity marketing.*
> *Bud Williams*

The University of Missouri once postulated that the cattle cycle was really caused by stocker operators getting into and out of cow-calf production and I think there is some validity to that theory.

In all margin trading activities where inventory is bought and sold, trading profits tend to be eventually bid into the base initial inventory.

This is because profits attract new players who bring in new money and they have to buy inventory to play. This new money was made elsewhere, is speculative in nature, and is not subject to the discipline of rational sell-buy market analysis.

In the spring of 2006 all commodity futures markets were bid to unrealistically high levels by "hedge funds" speculating on rising inflation predictions.

This speculation created record basis margins — the

difference between cash and futures — that made true hedging virtually impossible for real commodity producers.

Warren Buffett said when you realize you are playing poker with a lunatic the best thing to do is to leave the game. So, let's ignore the futures market and concentrate on the cash market.

But, first some basic economics.

Bud Williams said the ranching business is actually a three part whole.

You have grass.

You have livestock.

And you have cash.

Of these three, grass is always the centerpiece or primary business.

Bud said you don't buy or lease a ranch for its dirt but for the grass the dirt will grow.

You are, in effect, a grass farmer rather than a rancher.

However, you do not have to own livestock to sell your grass.

Custom grazing is an excellent alternative to ownership particularly during high calf price times because this is when the demand for it will be the greatest.

Grazing other peoples' cows on a custom basis pays a lot better than grazing owned commodity steers in high priced times for all of the previously presented math.

The cash flow is a lot better as well as grazing fees typically are paid monthly.

Marlene Moore, a long-time Bud Williams fan, told the Nebraska gathering that she found that adding a custom grazing component to her ranch gave her a lot more stocking rate flexibility than fully stocking the ranch with personally owned cows.

"In dry years I can call the cattle's owner and tell him to come and get them," she said.

Marlene uses Bud's sell-buy method with cows just like others do with stocker and feeder cattle.

In other words, when she sells a culled cow, she deducts this revenue from the cost of the incoming replacement cow for a truer picture of what the incoming cow is worth.

This helps to keep her feet planted firmly on the ground and helps her avoid being caught up in cow bidding fever.

What none of the Nebraska participants were comfortable with was that there appeared to be no way to get from high priced cattle to low priced cattle without a loss of net worth.

Bud's fast sell-buy inventory turn can provide a positive cash flow on the downhill slide but your net worth still declines.

This net worth decline can be particularly high for cow-calf producers because the heavier the animal, the greater the price decline in a falling market.

The government recognizes this risk and therefore allows you to depreciate purchased breeding stock to zero value over five years.

(Home raised cows cannot be depreciated but cash development costs can be deducted.)

Still most full-time ranchers don't use this depreciation allowance as a market price smoothing tool.

What they see is that they paid $1000 a piece for a cow and it is now worth $300.

To them this is a loss of $700 a head.

The only way to avoid this type of cowboy accounting loss is to sell out, ride the market down while staying in cash and then buy back in when prices hit bottom.

More on this in a moment.

First, let's take a look at the third piece of our ranch whole. Most of us never think about the cash portion of our farms and ranches.

One who does is Colorado sheep rancher Richard Parry.

Parry uses custom grazing and direct marketing to avoid the commodity business altogether. He is a graduate of Bud's schools and agrees with the three part whole concept.

While never working off the ranch, Parry has become an

accomplished investor of his ranch's cash. This has been a self-taught skill that he has learned through extensive reading.

"I was like most ranchers. Ranching was my life and I never considered off-ranch investments," he said.

> *A complex system that does not work is invariably found to have evolved from a simpler system that worked just fine.*
>
> *Murphy's Law of Computing*

"The big paradigm shift came for me in the drought of 2002 when I was forced to almost totally liquidate my sheep flock.

"This liquidation occurred at the top of the sheep market cycle and I suddenly had a lot of cash for the first time in my life.

"Rather than reinvest all of this back into high-priced sheep inventory, I decided to buy commercial real estate instead."

He bought apartment rental properties and mini-storage warehouses, which could at that time actually pay for themselves out of the rent.

He rode those properties up in price for two years and then liquidated his real estate holdings once they became overvalued.

"When rent will no longer buy the underlying property, you know it is overvalued," he said.

"Almost every investment has common-sense rules of thumb like this to tell you whether the market is over or undervalued. You just need to read enough to know what they are."

He then re-invested this real estate profit into natural gas properties just as the energy boom was taking off.

"The hardest thing to learn is that everything is cyclical. Nothing lasts forever.

"You have to be willing to sell over-valued investments and buy under-valued investments to create wealth.

"Buy and hold is a fool's game," Parry said.

And yet that is what most of us want to do.

Most of hate to sell livestock, real estate, or the stocks in our retirement accounts. We hate selling livestock so badly that most of us will only sell when we run out of grass.

Again, a quick market analysis shows this.

May, which is the end of winter annuals in the South and California, is, on average, the lowest priced month to sell feeder cattle.

October, which is the end of the grazing season for many perennial pastures in the West and Midwest is the second lowest month.

The best month to sell feeder cattle is August. This is when Gordon Hazard sells his cattle.

Selling in early August is no great loss because 85% of the season's gain on stocker cattle has been made by then in most of the country.

But, the grass is still green and we don't have to sell, so we don't.

Similarly, while freezing is the worst thing to do in a falling price market that is exactly what most of us do when the price of anything we own drops precipitously.

We become like deer trapped in the headlights of an oncoming Mack truck. We instinctively know we should do something but for whatever reason we don't.

Bud Williams warned to never build a strategy around doing something you won't do. Selling out and going to cash is just something that most of us won't do.

I know of several ranchers who cashed out and then bought back in when the market didn't decline as quickly as they thought it would.

Warren Buffett said he has never called a market peak correctly and has always cashed out years too soon.

Did I mention that Warren Buffett is, for all his poor market timing, the second richest man in the USA?

He said trying to pick the exact day of a market collapse would almost always guarantee your getting sucked down with it.

The rules of the game:

1. In a falling market, the worst thing you can do is to freeze and do nothing.
2. Last-out first-in accounting is recommended for all businesses with highly volatile inventory prices.
3. Everything has a cycle. Nothing lasts forever.

Good depreciation vs. bad depreciation

Bud Williams has pointed out that the farther you get from the abattoir door the less the price paid for the animal is tied to reality. A young cow is about as far as you can get.

It is true that the depreciation on high priced cows will help offset the profits from high calf prices, but this depreciation will be "real" because the cows will eventually actually decline in value.

This is what *Rich Dad/Poor Dad* author Robert Kiyosaki calls "bad depreciation."

However, there is such a thing as "good" depreciation.

Back in the 1970s, you could buy a tractor, depreciate it, and sell it for more than you paid for it. That's an example of good depreciation.

For cow depreciation to be good depreciation, the cow's value must increase, or at least hold steady while you depreciate her.

The $600 cow bought in 2001 and sold for $900 in 2006 is a good example that this does indeed happen.

Good depreciation produces what Kiyosaki calls "phantom income."

In other words, the cow's five year tax depreciation shelters most or all of the taxable income from the sale of her calves while holding her own value.

Of course, for any tax shelter to be of value your ranch must be producing taxable income.

So, what are the alternatives to buying more cows?

Robert Kiyosaki said the asset that produced the profit

always has first claim. In other words, it's okay to milk a cash cow but not to starve it.

Unless you are going to custom grazing or selling out, livestock sold must be replaced. Soil minerals and pH must be kept at their optimum, fences must be kept sound and water lines have to work. Bulls have to be replaced.

In other words, Kiyosaki said your first investment priority is to protect what you've already got. Only then can you think about what to do with the rest.

While I am a firm believer that you shouldn't pay one red cent more taxes than you legally owe, I am an equally firm believer that managing a farm or ranch so as to minimize taxes rather than maximize income is a fool's game. Usually tax avoidance means we are deferring taxes rather than escaping them. In other words, we are delaying a small problem today by creating a much larger one in the future.

The old Investment Tax Credit was the primary cause of the 1980s farm crisis because it encouraged farmers to take on capital investments they didn't need just to avoid taxes. Thankfully, we don't have this farm and ranch killer anymore but we still have a lot of farmers and ranchers who don't understand that wealth is created with after-tax dollars.

If you want to own land, you must pay taxes because land — as a non-depreciable investment - can only be paid for with after-tax dollars. This is also true of a home, stocks, bonds, gold, diamonds or any other non-depreciable "investment."

Having admitted my bias against tax avoidance schemes, long-term fertilization programs are a good income smoothing program. Fertilizer is deductible in the year in which the expense is incurred but long-term fertilizers such as lime and reactive rock phosphate have benefits that can last for many years.

By fertilizing heavily with lime and phosphate during the high-profit years of the cycle, you create a before-tax fertility reserve in your soil that can be drawn upon when times get tough and cash flow is low.

Ignoring tax minimization, there are better places to put your excess cash flow than others. Let's look at some of those.

Taking out a short-term livestock production loan at the bottom of the market makes good economic sense because the increase in the value of what you buy will more than pay for the loan. However, not paying short-term debt off at the top of the market doesn't.

The best use of after-tax dollars is to pay down short-term production debt. If you can stop paying your bank ten percent and start them paying you five percent, that's a 15 percent change in your cash flow. While this may not meet Warren Buffett's definition of a 15 percent return on your money, it will spend like it.

> *The rules of financial holes:*
> *1. First stop digging.*
> *2. It doesn't matter how comfortable your hole is, get out of it.*
> *3. Deduct the costs of digging the hole.*
> *4. Fill in the hole.*
>
> *Pita Alexander*
> *New Zealand Farm Consultant*

Paying down long-term real estate debt is not as easy a decision. In fact, expanding your long-term real estate debt may actually make good economic sense at the top of the cattle market.

The secret to good bank relations is to only borrow money when you don't need it. Long-term loans allow you to do this. Therefore, good times such as these are an excellent time to refinance your existing long-term debt.

After paying down short-term debt, the next best use of after-tax dollars is create a reliable supply of stockwater. Usually in a drought a ranch's stockwater resources run out long before its forage does. One of the highest risks in ranching is being forced to sell when you don't want to. This is only slightly more costly than not being able to buy when you want to. Both of these risks are related to stockwater supply.

Gordon Hazard, who self-describes himself as "tight but

not cheap" redid all of his ranch's stockwater ponds to make them more reliable in drought years. Anything that Gordon is willing to spend money on has to be important.

Unfortunately, pond construction is neither deductible nor depreciable. However, I know one grazier who charged his pond deepening expense to fertilizer as he spread the accumulated bottom muck on his pastures. You are much more likely to be able to get away with this as a full-time farmer or rancher than as a part-timer. Part-time ranches are under the microscope at the IRS and really need to be careful.

A good stock pond (tank) should be dug deep with a minimum of surface area as this minimizes evaporation. Ideally, the water should be piped to a stock tank below the pond rather than allowing the cattle to drink (swim, urinate, defecate) directly from the pond. Watering stock from streams is likely to increasingly bring unwanted attention from regulators due to the Clean Water Act.

Nothing is more threatening to your wealth than a perimeter fence that won't hold stock. I well remember the sick feeling in my stomach when driving home from church one night I realized that all the blue lights and ambulances blocking the highway were because my family's cows had gotten out.

Luckily this was in the era when accidents were considered accidents and not grounds for lawsuits. We don't live in that era today. Good fences not only make for good neighbors but are good liability insurance as well.

A fence that will hold both sheep and cattle will give you a lot more options than a cattle-only one and will cost only slightly more. An electric offset on the inside of the fence will give you power for temporary fencing and keep the stock off the fence, which will make it last much longer.

The New Zealanders say the regard a grazier has for his exterior fences matches the regard with which he holds his pastures. One Kiwi tourist told me that in a coast-to-coast tour of America he didn't see one good exterior fence. Ouch!

Every year is marred with a drought of some duration.

For graziers in high value enterprises such as dairy and beef or lamb finishing, small-scale irrigation could be a very valuable investment and it is depreciable.

In the arid and semi-arid Southwest and West where multiple-year droughts are commonplace, the development of non-weather dependent income sources should be a priority. One of the best is hunting lease income.

Keep in mind that it is just as easy to over-capitalize these ancillary income opportunities as it is a cattle operation. A per acre lease program where you have to make no improvements and provide no management is usually more profitable than trying to maximize hunting income through active management.

> *A business designed to be low cost doesn't need huge sales. It is always dramatically more profitable as long as it stays dramatically lower cost.*
>
> *Adrian Slyotsky*
> *The Art of Profitability*

A hunting lease provides passive income. A hunting program is active income. There's a big difference between the two.

Active income comes from your own active management. These are those activities we describe as "management-intensive."

Passive income comes from other people's active management. This includes rents, royalties and dividends from passively held stock or businesses.

A ranch can be shifted from active to passive income by hiring a ranch management professional. A dairy can become passive income under a sharemilker arrangement.

It may be better to get concession rent and let someone else develop a trail ride, or chuck wagon cookout on your ranch rather than attempt to do it yourself if you have your hands full with what you are already doing.

The primary problem with active income is that it is directly tied to your time. Any increase in leisure time has a direct negative impact on income. This is not the case with passive income. The "idle rich" can be idle because all of their income is from passive sources.

Passive income sources also offer insurance against "life accidents" such as disease and disability. If your horse fell on you tomorrow, what would it do to your family's income?

While the return on investment from passive income is usually much lower than what you can take out of a personally managed business, it is not tied to your own health or energy. This is why people typically seek to replace active income with passive income as they age.

In conclusion, you have good options available at the top of the market that will not be there at the bottom. Don't let these options slip by you without thinking about them long and hard.

The rules of the game:

1. Unless you are going to custom grazing or selling out, livestock sold must be replaced.
2. Good times are excellent times to refinance your existing long-term debt.
3. After paying down short-term debt, the next best use of after-tax dollars is to create a reliable supply of stockwater.
4. A fence that will hold sheep and cattle will give you a lot more options than a cattle-only one.
5. Small-scale irrigation can be a valuable investment and is depreciable.

Crisis management

Bud Williams said every grazier should have a "crisis" marketing plan. Even if there is never another case of BSE or Foot and Mouth, this plan will serve as a good template for action in a major drought or other market emergency.

Bud recommended that before doing anything else when you hear bad news is to make yourself a cup of coffee, calm down and think. He said that in every market emergency you normally have two to three weeks to act before the full impact sinks in to most producers and the livestock market.

He said it is absolutely imperative that you not join your neighbors in a "tough it out" mindset during this brief grace period. This brave "stiff upper lip" will eventually give way to panic, and people will do really stupid things in a panic.

Step One

The first thing you need to do is build a feed inventory. How much grass and stored forage do you have on hand? Do you have enough to see you through the crisis?

"In a drought," Bud said, "grass is more valuable than cattle."

Step Two

The second inventory to check is your cash reserves. In a market crisis, Bud said having lots of cash is the ultimate unfair advantage.

He said a ranch was made up of three inventories. These are feed (pasture), money and livestock. The least important of these inventories is livestock.

"Few ranchers ever went broke having too much grass and money on hand," he said.

"If you have lots of grass and money, you can always buy livestock when the crisis is over."

Step Three

Last, do a livestock inventory. Remember, the animals that will be most impacted in a crisis are those closest to slaughter.

Therefore, you should move to sell cattle closest to slaughter at whatever price they will bring as quickly as possible and buy back lighter cattle.

"Buying light calves gives you time. In a crisis, time is the most valuable commodity," Bud said.

The lighter cattle also eat less grass and this can increase your feed inventory.

He said owning cattle through the feedlot was always too risky to even consider without a forward contract or hedge.

"The problem with the feedlot is that once the cattle are on feed the only way out is through the abattoir door. You don't have any marketing flexibility in a feedlot."

He said rancher fears that the person on the other side of the futures contract would go bankrupt and leave the hedger exposed were unfounded.

"The full faith and credit of the entire exchange stands behind the contract, not just the net worth of the trader. Keep in mind, cattle are a tiny amount of the futures market and a cattle crisis would not have any major effect on the Chicago Mercantile Exchange."

He said that allowing feeder cattle to get over 800 pounds was extremely risky because your marketing options start becoming limited. He said lighter cattle always take less of a hit in any crisis because they are farther in time from the ultimate reality of the abattoir.

Bud said a policy of selling feeder cattle at less than 800 pounds and buying back lighter calves was always less risky than trying to maximize the weight gain on one set of feeders.

While most ranchers don't like to consider it, he said after heavy feeders the first category of animals you should consider selling are your mother cows. This is because they eat the most grass and are frequently the most over-priced class of livestock.

During the Canadian BSE outbreak, a Canadian attendee at the Stockman Grass Farmer's Marketing School said Bud was right and that young cow prices didn't fall at all for two to three weeks.

He said the biggest impact had indeed been on fat cattle and older cows. In other words, the two classes nearest to slaughter.

> *The only thing we have to fear is the lack of fear itself.*
>
> *Larry Summers*
> *US Treasury Secretary*

Because of the shortage of cow abattoirs in Canada, the market for older cows largely vanished. By late September 2003, some Canadian dairymen were actually having to pay to have their cull cows disposed of.

Bud said that if there continued to be no slaughter market for older cows in Canada ranchers should consider shooting them rather than continuing to hold them in hopes of a border opening or a government bailout.

"Unprofitable ranchers will eventually quit taking care of their cattle. I saw many ranchers let their cattle and sheep just starve to death during an Australian drought. They would have been many dollars ahead to have just shot them and saved their grass."

He said that financially successful ranchers all learned to love their grass more than their cows.

"The primary reason ranchers get hurt in a drought or market crisis is that they absolutely refuse to sell their cows," he said.

"The best strategy in any feed emergency is to sell your cows and keep your heifer calves. This gives you genetic continuity but tremendously lowers your feed demand and creates a surplus inventory of cash."

> *Be not afraid of growing slowly, be afraid only of standing still.*
> *Chinese proverb*

He said holding heavy or grown cattle for better prices is always the wrong decision.

"Your first concern is to preserve your feed supply. You lose the least money in a crisis when you sell cattle fast.

"Do something drastic to reduce your numbers and conserve grass."

Bud Williams said that every animal on your farm should have a price tag on it and if someone is willing to pay the price you need to let the animal go.

"The ability to sell cattle well is the ultimate unfair advantage," he said. "Every day that you don't you sell your cattle you have, in effect, bought them. If you aren't buying more cattle because you think they are priced too high, then you should be selling those you already have."

Unfortunately, Bud said, most people in the cattle business do not like to sell and prefer a "buy and hold" type of marketing whereby cattle are only sold when the sale is forced by the weather or economic circumstances.

"We buy when we want to but only sell when we have to," he said.

Bud thinks the reason for this is the fear on the part of most producers that their neighbors might get a penny more than they got.

"If you want to compare numbers with your neighbors, compare profits not sales prices. I guarantee you that the man who sells cattle frequently at less than the best price will always be ahead of someone who tries to make a killing on each animal."

Bud said that cash flow is only generated through the

sale of animals. While a "buy and hold" strategy can maximize capital gains in a rising market, it maximizes capital losses in a falling one. Bud said the worst thing you can do in a falling market is to freeze and ride it all the way to the bottom.

"The faster you sell and buy back even cheaper replacements, the better off you are."

Apparently, Wall Street traders do not like to sell their holdings anymore than ranchers. As a result all Wall Street trading houses have their traders' computers wired to automatically sell their entire inventory of a stock if it falls beyond a certain point.

In his book *Deflation*, investment adviser, A. Gary Shilling, explained why the trading houses have done this.

He said it is all a game of math and percentages. To recover from a 50 percent decline in the value of something requires that it increase in value by 100 percent just to get back to where it was. In other words, to return to where you were after a fall takes twice as much increase in price.

For example, Shilling points out that if one had been fully invested in the American stock market from 1946 to 1991 and had missed the 50 weakest months, it would have improved your return by 70 percent over being fully invested at all times.

Conversely, if you had also missed out on the 50 best months in the same time period, it would have only lowered your return by less than a half a percent.

Shilling said this last point is important because it emphasizes that trying to hold out for the last penny before getting out of a bull market is a fool's game. The risk from riding the asset down in price far outweighs missing a few dollars from the final blow-off before the wreck.

Shilling said the other serendipitous part of selling at the top of a market is that one has a lot of money to buy back in with at the bottom. For all of these reasons, Shilling said a market timing approach focusing on avoiding the big price drops is by far the best strategy.

The rules of the game:

1. In a crisis, first build a feed inventory.
2. Check your cash reserves.
3. Sell cattle closest to slaughter at whatever price they will bring as quickly as possible and buy back lighter cattle.
4. Mother cows are the first category of animals you should consider selling.

Get the hay out and the profits will follow

If you compared the return on assets of a well-managed cow-calf ranch in the upper Midwest to America's 500 top industrial corporations, who do you think would come out ahead? The answer will probably surprise you.

According to a financial analysis of several hundred Midwestern ranches by Dr. Barry Dunne when he was at South Dakota State University, well-managed cow-calf operations in his area showed a 10 year average return on assets of 18.6 percent. This compares to six to eight percent average return by the top industrial corporations in the S&P 500 stock market index.

"What we are using here is managerial accounting," Dunne explained. "The companies that are valued by Standard and Poors don't revalue their factories at market value. They use cost and the return levels they report are based on cost and they do accrual adjustments when they sell assets the same way we do. This is probably the closest comparison to the way other businesses are evaluated that agriculture has ever seen."

Unfortunately, this high profit group makes up only 16 percent of the total group of ranches analyzed. Most cattle ranchers showed far poorer returns than the top group. Sixty eight percent of ranchers showed only a 2.9 percent return and the bottom 16 percent had a negative return of 15.5 percent.

"What was significant to me was that here were a sizable group of people in the same region, with the same market prices, the same problems and opportunities and yet there is a huge difference in profitability between them. That difference is management."

Dunne said the statistical analysis of 240 ranches in (mostly) South Dakota and Nebraska over a 10 year period was done for his doctoral thesis. The analysis used the Standard Performance Analysis figures (SPA) that many universities have been collecting on cooperating producers for the last decade.

Dunne said care was taken to not include any seedstock ranches that might skew the return. The average size ranch studied was 500 cows but the ranches themselves ranged from as small as 20 cows to nearly 5000 cows. The average age of the ranchers studied was 45.

He said actual costs were used and machinery was depreciated at its normal expected service life rather than at the much faster tax depreciation rates. Purchased cows were allowed to be carried at zero cost after five years but raised cows had to carry a cost ($600 to $700 a cow) to reflect the cost of raising them. When anything was sold it was marked up to market price by accrual.

"I believe that for anyone to fully understand ranching you must have three sets of books. One for the IRS. One for the Bank and one for you.

"For the IRS you want to take advantage of all of the fast depreciation the law allows to shelter family income. For the bank you want to mark up all of your assets to current market value to increase your collateral. For your own set of books you are primarily interested in actual cash flow rather than asset value changes and depreciation."

Dunne said in his analysis he was primarily interested in determining the effect of management on the profitability of a commodity-priced ranch. He said his analysis found that many things being taught as "good management" today actually had little to no positive effect on profitability.

"The only production measure that had any significant impact upon profitability in a cow- calf ranch was reproductive rate," he said.

"Weaning weights, average daily gain, stocking rate - none of these had a significant impact upon profitability. The high profit producers produced four percent more calves than the average producer. The secret (to profitability) is more calves — not heavier calves."

Equally surprising return on assets was herd size neutral.

"We did not find any point of diminishing returns in the smallness of a cow-calf operation. Well-managed small cow-calf ranches showed the same rate of return as large well-managed cow-calf ranches.

"Economies of scale can work both ways. Often smaller ranches have more access to under-valued resources than larger ranches. Just being big is definitely not the answer (to ranch profitability)."

Location didn't matter either.

Ranches in dry areas showed no significant decrease in profitability compared to ranches in higher rainfall areas.

"Management ability tends to wash out all perceived (climatic) advantages and disadvantages. It's really amazing when you think about it."

Dunne said making an above average profit was not any one thing but a complex set of relationships involving investment, production, marketing and annual (cow) cost.

"High profit producers are able to find value, invest below the market, yet not suffer in major production efficiency loss and actually have higher reproduction rates (than the high cost producers)."

For example, high profit producers showed a much lower investment per cow than the average. The high profit producers had a total investment per cow of $1400 compared to $2300 for the average rancher. This investment includes the cost of the cow and a portion of the bull, land, machinery - everything but direct cash expenses.

"The study confirms that if it rusts, rots or depreciates you don't want to put your money into it," he said.

While machinery was a big part of these investment costs, paddock fencing and water developments can also run up investment per cow.

Time of calving can also have a big influence on the level of investment per cow. He said his own family had invested thousands of dollars in calving barns and windbreaks that would have been unnecessary if they had been summer calving.

> *Productivity is about the relationship between quantity of output versus the quantity of input. It's a quantity measure, not a profitability measure. You can be highly productive doing absolutely the wrong thing.*
>
> *Jason Jennings*
> *Less is More*

"An Australian visitor once told us that he didn't have any blizzards where he was from, but if he did, he sure wouldn't be calving in them. Unfortunately, such common sense has been sorely lacking in (North American) ranching."

While most ranchers think that as far as land ownership is concerned one has to "marry it or inherit it" to have a low land cost per cow, Dunne's analysis does not confirm this.

"Most of these (high return producers) are not fifth generation ranches who inherited their land. They are people who were willing to buy when everyone else was selling. There have been times when you could buy ranch land at extremely reasonable prices in the last 20 years," he said.

Also, he pointed out that a return on assets analysis does not favor inheritors. The more of your equity you have trapped in land the harder it is to show an above average return on assets.

Average direct cash costs per cow were also dramatically lower for the high profit group compared to the average rancher. The high profit group had an annual cash cost per cow of $250 versus $400 for the average rancher. (The national average is $401.) A major difference between the high return producers and the average was lower hay costs and lower veterinary bills.

He said to keep in mind it was possible to have a high return on assets and a low income as long as you have a low investment.

However, he said the reverse of this was also true. You can have high costs and a high return if you are an exceptional manager.

"We didn't find a single recipe for profitability. We found a set of relationships," Dunne said. "It is not the kiss of death to have above average cow costs. It is the kiss of death to have above average cow costs and below average levels of production."

Dunne said that he personally would prefer to go the low cost route but that there were a few very viable high cost/high return producers in their study. He said high cost producer people have to be exceptional managers to make a profit.

"I cannot over-emphasize that your initial investment decisions are critical. If you have a big investment, you've really got to have a production system and a marketing system that is just absolutely error-free."

More typically, he said there was a tendency for ranchers with a high level of investment per cow to dig themselves ever deeper into the over-investment hole.

"The level of investment is the whole thing (that drives management). If you invest too much you start feeling the pressure to have high levels of production. There's a tendency to overapply technology and overspend.

He said the necessity to keep machinery and bull costs low was not well-received by the majority of ranchers.

"Eighty-four percent of our ranchers just don't get it. They don't want to be told they have made bad decisions.

"Too much machinery drives up both direct costs and annual repair costs. If you own it and it breaks do you not fix it? Of course you do and those kinds of costs tend to pyramid."

He said one rancher in the survey has a ROA of over 20

percent. However, his newest tractor is a 1969 model and yet it looks and runs like new because he changes the oil and keeps it well maintained.

"We haven't paid enough attention to the little things that really count."

Genetics are another source of overspending.

"The average rancher doesn't want to hear that a $5000 bull is not any better than a good $1500 bull. And yet it is just those kinds of decisions that absolutely are (financially) killing them. They spend too much money on things that don't really matter."

The Commodity Cycle
1. Everyone cuts costs.
2. Everyone cuts prices.
3. Everyone cuts profits.
Thomas A. Stewart
The Wealth of Knowledge

He said perhaps the following analogy would put the importance of a high rate of return into perspective. "If you have an 18 percent return to land, labor and management, you can draw out $35,000 for family living expenses, pay off all of your debt in 10 years, and do this with only 202 cows.

"If you are an average producer, that same $35,000 in family living expense will require 972 cows and you'll never be able to pay off a dollar of debt."

Dunne said for high management ranchers the number of cows needed was determined by the family income goal. "If you want a $75,000 annual income, it is going to require around 400 cows. If you want a $100,000 annual income it is going to take around a 130 more."

He said to keep in mind these figures were not after-tax. He said ranching had enough tax advantages that most of your annual income can be tax sheltered.

"I think the continual preaching of the necessity to get bigger is a big con. A lot of older ranchers thought they would have less competition for resources if they ran all of the young people out of town (with their negative viewpoints). Now, those towns are dying and there's no one there to allow them to cash out of their ranch investments.

"Heck, I would rather have a lot of 200 cow ranches and a thriving community than a few 1000 cow ones and no community. Our data shows that a ranch can be small and yet very efficient and make a lot of money."

The rules of the game:

1. The only production measure that has any significant impact upon profitability in a cow-calf ranch is reproductive rate.
2. Don't sink your money into things that rust, rot, or depreciate.

Ways to increase return on assets:

1. Rent or lease land rather than buy it.
2. Occasionally refinance land to lower equity.
3. Buy hay rather than produce it.
4. Buy replacements (which can be depreciated to a zero value in five years) rather than produce them.
5. Keep fully depreciated cows in the herd as long as they produce a calf.
6. Use fully depreciated machinery and vehicles.
7. Buy breeding stock at the bottom of the cattle cycle.
8. Keep bull purchase costs reasonable.

Concentrate on things you can change

"What you paid for an animal six months ago is irrelevant because you can't change it. If you can't change it, you can't manage it, so forget about it," Bud Williams said.

Similarly, if you think you can predict tomorrow's livestock prices, Bud said you need to be trading futures and not raising cattle. He said the futures markets will return a thousand times more profit to correct speculation than cattle ownership will.

"To make a living ranching you have to put the majority of your mental energy into managing what can be managed - which is the price relationships between classes and weights of livestock on today's market. In other words, forget the past, ignore the future and pay attention to right now," Bud said.

This is excellent advice. However, as the general price level of livestock changes, you may need to re-balance your inventories.

For example, the time to build your cash inventory is when livestock prices are high and the time to build your livestock inventory is when livestock prices are low.

In other words, prices are irrelevant in terms of on-going operations but not irrelevant in terms of livestock expansion.

Bud said the problem with most ranchers is that they are always getting out of precisely what they should be getting into and vice versa.

Rather than taking advantage of high prices in high-priced times to build cash reserves, many ranchers should forego excellent prices and build brood cow numbers.

Beef production today has become as much a matter of harvest weight as cow numbers.

Because of our shift to later-maturing cattle genetics, every year we produce more and more salable beef from fewer and fewer cows. This makes an increasing cow herd particularly worrisome.

Just as circumstances have combined to produce the highest prices in history these same circumstances could eventually result in, let's just say, significantly different prices in the future.

> *Take profits! Pay taxes! If you don't take profits, you won't have profits and the tax man will be the least of your worries.*
>
> *Jim Cramer*

Old-timers will remember that the high prices of 1973 triggered an overexpansion that resulted in an almost total economic collapse in American agriculture just eight years later.

I was in Brazil in August 2003 and there were literally beef cows everywhere. It reminded me of the Southeast in the early 1970s.

Brazil already has the second largest cowherd after China's and is growing it as fast as they can. These cows are going to be a major factor for all beef producers in the future. So, don't forget about them.

As America's second richest man, Warren Buffett, said, "The secret to success is in being able to keep your head when everyone else is losing theirs."

Buffett, by the way, in 2006 was using his many companies' profits to build cash reserves rather than expanding.

Note that he was not liquidating his holdings. He just temporarily stopped his expansion and built up cash. He will use this cash to expand when the inevitable "return to value" occurs.

This is the key difference between a crapshoot and a

business.

Businesses are not built to be sold whenever the price of scrap gets high. A business should be managed so as to make it last forever.

Part of that long-term management approach is being very careful with what you do with your profits. Reinvesting excess profits back into what just made them is almost always the wrong thing to do.

So what should you do?

While staying with Vaughan and Auriel Jones in New Zealand, I was rummaging through Vaughan's enormous reading pile and came across a newsletter by a New Zealand financial farm consultant named Pita Alexander that addressed this re-investment issue.

Here's Pita's advice about what to do whenever you find yourself at the top of the price Ferris Wheel.

1. Take Lots of Pictures.

It could be a long, long time before such prices will be seen again. So long, in fact, that without photos we may forget that such a time ever existed.

2. Catch up on key deferred maintenance.

Use today's profits for things that lower costs long-term.

Legumes are the primary production cost-reducer in pastoral agriculture. They improve both reproductive rate and average daily gain. Legumes are very sensitive to soil calcium and phosphate. Now is the time to get your soil calcium and phosphate levels up and in balance.

Fertilizer is also tax deductible and can "smooth" your tax liabilities.

3. Look for short-term opportunities with a quick return.

The top of the market is not the time to be making long-term purchases such as brood stock. Look for situations where purchased inventory can turn in 100 days or less.

You need to always keep 20% of your farm's annual gross income in cash.

4. Reduce interest-bearing debt.

Paying down debt is a better use of your money than borrowing and expanding at the top of the market.

Borrowing against cyclically high collateral is extremely dangerous because the liquidation of the collateral may not pay off the debt if there is a precipitous fall in its underlying value.

In other words, when you leverage your bet in a high market you can wind up losing more than you bet because your pledged collateral won't cover the loan. Frequently, the only way out of such a situation is personal bankruptcy.

5. Avoid buying depreciable assets as a tax dodge.

Research in the USA has found that 95% of all depreciation is "real." In other words, the item you buy actually does lose the value depreciated. You are far better off to pay the taxes.

6. Never make a capital investment if your heart is not in it.

The odds of your being able to recoup all of a capital investment through subsequent liquidation is remote. Also, most farm capital investments add nothing to the appraised value of the farm.

Capital investments are only cost-effective when they allow you to do more work. Make sure you like the work.

7. Face the truth about today's farm real estate prices.

If you can afford to buy a farm that will support you with no outside income, then you probably have enough money to retire and not work at all.

Amen to all of Pita Alexander's top-of-the-market

recommendations. I also agree with his observation in another newsletter that the era when the "average" farmer could ever make a profit is over.

He defined the average as where "the worst of the best meets the cream of the crap." He said the top 10% of farmers do better and better regardless of the prices and the bottom 10% do worse and worse, regardless of the prices.

As Bud Williams said, "Management always trumps prices."

While the top of the market may not be a good time to grow gross production, you should never stop trying to grow your bottom line.

Alexander said the difference between the winners and the losers is that the winners understand you have to spend money to make money.

He said the average commodity producer gets so wrapped up in cost-control that he ignores opportunities.

"You cannot grow a farming business by only shrinking the expenses. You must be willing to spend $50 to make $100."

Following Nature's model

All land-based agricultural production systems exist to capture solar energy in some form that can be converted into human energy. These enterprises all measure their profits on a land unit acre basis.

A beef feedlot, confinement chicken or hog house is not a land-based agricultural production system because the energy is captured on land elsewhere and brought to the animal.

Such feed importing enterprises typically do not measure their profitability on a per acre basis but on a per animal basis.

This difference may sound subtle but it is not. Here's why.

Of all the world's energy sources, human energy is the most expensive and solar energy the least expensive.

Therefore, production margin per acre is the widest when solar energy use is maximized and human energy is minimized.

Industrial agriculture has done this by substituting petroleum energy for human energy. However, the widest of all per acre margins occurs when an animal harvests its own feed directly from the land where it is grown.

With fruits and vegetables this can be a pick-your-own business model, but in animal agriculture this means direct-grazing.

Everything and anything that intervenes in the capture and conversion process diminishes the return per acre. If you can grasp this simple fact, you are 90% on your way to making a high return from agriculture.

If we study Nature's production model, we find that animals find it energy conserving to not compete directly with other animal species for their food supply. As a result, they develop specific feed niches.

Among ruminants, the four major feed niches are coarse grass, fine leaf grass, broadleaf weeds and browse.

These four feed resources are found in all natural temperate environments but vary as to the percentage mixture. This variation is primarily a function of the reliability of rainfall but also is impacted by slope, aspect and periodic flooding and fire.

The dominant feed mix determined which species was the dominant grazer.

For example, grass-niche bison tended to be a dominant species in the mostly grass North American Plains whereas deer tended to dominate in the browse-rich East and Southwest.

However, there were deer in the Plains and Bison in the East because there were still the four feed niches - just in differing amounts.

The point to remember is that nowhere in Nature was there only one species of grazing animal. There was always a mixture.

Africa has the highest natural meat production per acre in the world because it has the most diverse mix of grazing animals.

Think of the African giraffe browsing the tops of the trees. You don't see that kind of species diversity in North America or Europe and so the harvest efficiency is lower.

The bottom line is that multiple species grazing always increases the economic harvest from a land area. The key is that the dominant species should match the dominant forage group.

Increasing net return in agriculture comes primarily from increasing the solar interception interface (pasture density) and the efficiency of harvest (MiG and multiple species).

Graziers with only cattle do not see broadleaf weeds and brush encroachment as an economic resource but as a problem to be eliminated.

Add sheep and goats and the problem becomes a source of cash flow.

Conversely, a prepared seedbed sward of annual ryegrass, offers little opportunity for multiple-species grazing because there is no feed diversity.

Holistic economist, Peter Senge, said we must realize that "we and our problems are always a part of the same system of thinking."

> *The Japanese believe that growing a business is like growing a tree. It takes 25 to 50 years.*
>
> *Thomas A. Stewart*
> *The Wealth of Knowledge*

The system of thinking that creates most of our problems in grassland agriculture is that we try to impose "human centric" policies upon Nature. By human centric, I primarily mean the eternal human desire to simplify things so as to eliminate the need to think.

At the heart of industrial agriculture has been a "dumbing down" of how Nature works. For most modern farmers, weed control can only come from a herbicide. Internal parasite control can only come from a wormer.

Winter feed only comes from a large roundbale or the silage wagon.

And, soil fertility only comes from a fertilizer sack.

All of these are easy solutions that require no thinking. Just bring your checkbook.

The problem with this approach is that it narrows the monetary margin per unit of production and therefore requires an increase in volume to offset the loss in knowledge.

As organic farmers show us, there is always a "natural" solution to every problem in Nature. The problem is that this "natural way" always requires more knowledge, forethought and pre-emptive action.

In other words, if your goal is to maximize margin per unit of production you must primarily work on preventing problems rather than trying to solve them. To prevent problems you must understand the root cause of the problem and then avoid that situation.

That requires knowledge and knowledge requires learning time. In other words, obtaining knowledge is never "easy" because it always requires time.

All natural plants contain toxins known as "anti-quality" factors. These toxins help prevent that plant from being totally removed by its grazing "predators" such an insects and ruminants.

But, these anti-quality elements help prevent internal parasitism in ruminants and are also thought to help prevent cancer and other maladies.

Unfortunately, agronomists have been breeding these anti-quality factors out of forages to increase animal intake on man-made monoculture pastures.

As your doctor can point out, the only difference between a medicine and a poison is the dose. Grazing animals apparently dilute these plant toxins by grazing on many different plants which they can't do on a monoculture.

For example, goats - which always eat sparingly of any one species at a time - can eat plants too toxic for cattle and sheep.

What causes "problem" plants such as Leafy spurge is the lack of grazer diversity. Every plant is a part of some grazer's feed niche.

While we tend to think of cattle as a bulk-grass eater, sheep as a fine grass and forb-eater and goats as a browser, all three species can and do eat some of all four major forage groups.

European cattle, for example, evolved in a forested environment and relish mineral-rich broadleaf tree leaves. Similarly, goats will also eat some young grass.

That there is some cross-over does not invalidate that

each species is primarily seeking a particular feed niche. This was because the feed competitors were always close at hand.

If we study Nature, we find that most grazers grazed in close proximity to one another as this maximized protection from predators. However, this co-grazing also greatly helped in controlling species-specific diseases.

From a health standpoint, the most dangerous thing to do is to get cooped up in close proximity with a bunch of your own species. In other words, it is healthier to live in a house with your dog than with another human.

Most animal health problems, including animal parasitism, are species specific. Mixing grazing species helps to provide a natural "firewall" to contagious diseases and parasites.

For example, cattle are a dead-end host for sheep and goat internal parasites and vice-versa.

However, the pig is even more efficient at eliminating internal parasitism in ruminants as they concentrate on eating fresh manure seconds after it hits the ground.

Unhealthy animals are not "natural." Health is what is natural.

The rules of the game:

1.To prevent problems, you must understand the root cause of the problem and the avoid that situation.
2. In Nature there is always a mixture of grazing species.
3. The dominant specie should match the dominant forage group.
4. Mixing grazing species helps to provide a natural "firewall" to contagious diseases and parasites.

Enterprises, holons and centerpieces

Joel Salatin does a masterful job of managing 13 income producing enterprises on his 500-acre Virginia mountainside farm. With only 100 open acres, in 2002 Joel's farm produced $400,000 in gross income and provided four full-time "white collar salary" jobs. The farm has subsequently grown to around 500 open acres in 2006 through leasing.

Joel said there are three primary ways to increase gross income from a farm. One is to add irrigation. One is to direct market and the other is to stack enterprises. He has done all three.

However, he warned that new enterprises cannot be "stacked" willy-nilly. They must be carefully added as a puzzle piece that "fits" into the farm's "whole."

At the heart of a farm must be a "centerpiece"enterprise. He chose pastured broilers because he liked raising chickens and because it provided the frequent cash flow needed when the farm is your only income.

Joel pointed out that with the exception of his initial pastured broiler operation all of his subsequent enterprises started out as "holons" that organically grew into "enterprises." He also said there have been several holons that looked promising but were discarded because they didn't "fit."

Holons have been defined as "independent dependencies." In contrast to an enterprise, which exists to pay its own way, a holon exists to strengthen a pre-existing proven enterprise. In other words, a holon pays its way primarily by the service it provides to a cash producing enterprise rather through its own independent production sales.

Joel said that beef cows were his first holon. He explained that chickens cannot make use of tall pasture. His first cows were not purchased to produce high dollar grass-finished beef but solely as an economical way of keeping the pasture short for his broilers. In other words the cows' grass growth control "service" was the only income Joel initially sought from the cows.

Over time, Joel said that while still performing "service" to the broilers the beef cattle grew into an income producing "enterprise" in their own right and soon had their own satellite holon.

This holon was a rolling hen house he named the "eggmobile" that followed the cattle through their pasture rotation to scratch up the cowpies and help control flies and internal parasitism in the cattle. He said the "profit" from an eggmobile does not come from the eggs it produces but from the pasture sanitation "service" it provides.

However, the eggs produced are also profitable in the traditional sense because much of the feed required by such free-ranging chickens comes from the bugs they control.

Similarly, the first pigs on Joel's farm were used as a low cost way to turn over and aerate the winter bedding manure pack and turn it into compost. Here again he was seeking a service and not a salable commodity.

Joel said the beauty of a holon is that it allows you to "test drive" a potential enterprise to see if it "fits" both your farm and your own prejudices and predilections. Since little new capital investment is required for a well-chosen holon, it allows you to fail inexpensively. He said several of his holons had not survived the "fit" test and had been discarded.

Because holons essentially turn "trash into cash" they have extremely wide production margins that can be deceptive and potentially misleading.

Joel warned that these wide margins per unit of production exist only as long as the feedstuff consumed by the holon is considered to be worthless or worth removing. This means that holons necessarily have to be run at very low stocking rates.

For example, once chicken numbers start to exceed the natural bug production of the pasture they require more of their own feed resource. At the point where the increased feed costs surpass the value of the service they are performing, the hens will cease to be a "holon" and become an independent "enterprise."

The first rule of holonic ranch development is "protect the Mother Ship." A holon has to enhance the primary enterprise's profitability - not decrease it - or it shouldn't exist.

Sheep can be a holon on a cattle ranch until there are no more broadleaf weeds to eat. Once the weeds are gone the sheep become competitors rather than complements to the cattle.

Meat goats also make an excellent holon for a cattle ranch. However, a major problem with meat goats is that they tend to rather quickly eat themselves out of their preferred brushy forage niche. Once goats have to eat with their head lower than their shoulders they cease being a "holon" and become a major parasitism-infested "enterprise" that competes with the cattle for the feed resource.

A few free-ranging pigs can help keep the cows parasite free by eating their manure and becoming a dead-end host for the cattle parasites. However, once pig numbers exceed the "free" manure feed source they switch from pasture sanitizers to pasture destroyers as they must root to feed themselves.

Joel's enterprises mix 100 percent grassfed beef with grain fed poultry and pork. However, all enterprises use pasture as their base. In this way, the bought-in grain becomes a major fertilizer input for the pasture while simultaneously increasing gross production.

He said it was important to only invest in capital items that had multiple uses to get more value out of them. For example, his tractor and hay wagons are used for logging in the winter.

Hoop barn structures that are used for sheep (later discontiuned) and layer shelters in the green season shelter hogs in the winter.

> *Measurement is the key to success. Figure out what works, and do it more. If you measure it, it will improve.*
>
> *Seth Godin*
> *The Purple Cow*

He said that by only buying capital items that have multiple uses he was able to produce a dollar of annual gross sales from only 50 cents of capital investment. This is in sharp contrast to the four-to-one ratio of the average American farm.

All grain, stocker calves, lambs, and feeder pigs are purchased locally and a local farmer also does the feed milling.

A cattle farmer neighbor who summered his cows on the mountain next to Joel drove his cows and calves down to Joel's corral. There the calves were sorted off, walked across Joel's personal electronic scale and sold without any hassle, huff or sale barn commission. The neighbor was happy and so was Joel.

This was so much better than the sale barn cattle he had been buying that he said he now buys all of his stocker calves directly from local farmers.

Joel said there is often more value in keeping something as a small holon than in growing it into a true enterprise. This is particularly true if you are not into direct marketing where you can get a premium price that offsets the low volume of production.

He said to remember the most valuable consumer of a holon's production is your own family. For example, Joel's family has chosen to only grow vegetables for their own consumption and not expand gardening into a for-profit enterprise. The first beef, eggs and pork the farm produced were also initially for the family's own consumption.

Often holons grow into enterprises even when they are not planned to do so.

As an example, the $5000 on-farm sawmill was initially purchased just to provide lumber for the farm's own uses. However, the sawmill grew into a major income producing enterprise when custom home builders began asking for specialty cut lumber and beams.

Joel said to remember the sequence for all direct marketed products.

First produce it for your own family.

Give some to your neighbors.

Sell it to your neighbor's friends who have asked for it.

Turn it into a business.

While starting holons can be as much play as serious business, choosing a centerpiece enterprise certainly isn't. This is the most important decision in a holonic financial structure.

The centerpiece enterprise forms the hub around which everything is built. If it collapses it takes everything else with it. Until you have a functioning profitable centerpiece, you shouldn't add any holons.

Here are some guidelines for choosing a centerpiece.

- The centerpiece should be chosen so as to maximize the inherent advantages of your climate, location and your own pre-existing skills.
- The centerpiece should provide frequent cash flow.
- The centerpiece should be able to make money ten years in ten regardless of commodity cycles. Margin enterprises or enterprises where you can control your sales price make good centerpieces.
- The centerpiece should be scalable. Adding one more unit of production should require little to no extra labor input.
- The centerpiece should be relatively weatherproof.
- The centerpiece should have a high return to land and labor.

Joel chose pastured broilers as his centerpiece enterprise

because of the frequent cash flow it produced. However, he was able to do this because of the tolerant attitude Virginia had toward on-farm slaughter houses. In states with less tolerant attitudes, using broilers as your centerpiece enterprise would be far too risky because on-farm slaughter is a big part of the enterprise's profitability.

He said that all livestock enterprises had three income centers. These are production, processing and marketing. He said it was his experience that each of these were about one-third of the profit.

Ruminant brood stock enterprises do not make good centerpieces due to their low return to the land due to high body maintenance costs and poor cash flow. However, they do make excellent holons.

Beef cows are highly profitable when the feedstuff they live on is worthless or worth removing. Just as Joel initially did, cows are most profitable when used as a tool - a means to an end rather than an end in and of themselves.

Because cows double their forage intake upon calving, they are excellent tools to use to manage forage surpluses in the late spring on land inaccessible to silage or hay machinery.

They are also an excellent pioneer enterprise to open up and condition an overgrown farm or ranch for subsequent higher value stockers, sheep or dairy cattle or just to improve the land's real estate value.

So, why are there 800,000 cow-calf producers and only a few thousand stocker, sheep and dairy producers? Because cows are also an excellent holon to the most common centerpiece - an off-farm job.

Cows do an excellent job of adding value to what would otherwise be leisure time for an employed person. There is very little on a cow-calf operation that has to be done in an extremely timely manner and this makes them excellent holons for a person with an off-farm job.

Remember, the purpose of a holon is to add value to time or feed resources that would otherwise be wasted. If you have no surplus time or feed, forget about adding holons.

As Joel pointed out in his speech, a full-time farmer or rancher is first and foremost selling his time. All "enterprises" must have a high return to the labor involved. This requires scalability.

Grazing animals are an excellent example of a scalable enterprise. It requires very little more time to open a gate for a thousand steers than it does for a dozen.

Greg Judy and Gordon Hazard both run over 1500 cattle as part-time enterprises. They are able to do this because they have designed their operations so that the animals do the work rather than them.

Joel Salatin said most of us prefer to copy others rather than come up with something original. He said this is fine as long as you are sure you are copying something that will fit your own goals.

For example, most large cow-calf ranches function primarily as investments rather than businesses. Investments have been defined as things that allow you to stay rich. In other words, they are primarily hedges against inflation.

> *The hardest thing in business is to keep the entreprenurial spirit alive and flourishing in the wake of continued success.*
>
> *John Ellis*

In contrast, businesses are things that make you rich. They do this by providing frequent cash flow.

Gregg Simonds of Ensign Ranches makes a pretty compelling case that on large acreage Western ranches wildlife should be the centerpiece enterprise rather than cattle.

"I know what I am going to get for our elk hunting next year. I don't know what I am going to get for the cattle," he said.

He said the ability to price the resource and its lessened vulnerability to drought made a wildlife program's income much more reliable than cattle. However, he said he wouldn't want to try to run a wildlife program without cattle.

With planned and controlled grazing, cattle can be used as tools to sculpt the landscape so that both the wildlife production and the ranch's real estate value is enhanced.

Similarly, Missouri grazier, Greg Judy, was able to obtain free or nearly free grazing leases because he knew how to create a beautiful landscape with grazing while simultaneously enhancing the land's hunting. To do the same thing with a hired-in rotary mower would cost the absentee land owners many thousands of dollars a year in out-of-pocket expenses.

In other words, Judy's land owners "save" thousands of dollars by "giving" Judy the land's grazing rights. These people aren't crazy but are financially shrewd. Vegetation control is in many instances the major "value" of ruminant brood livestock and should be sold as such.

He said all "wild" properties should start with cows and/or horses and not stocker cattle. Stocker cattle should come only when the forage resource has been improved to the point where the cows become overly fattened.

In conclusion, trying to start a multiple enterprise farm or ranch all at once is an extremely frustrating experience and one that could be hazardous to your financial health. If you currently have an off-farm job, start with a holon that will add value to your time and provide you with a learning experience, and let the farm or ranch grow "organically" from there.

If you are a full-time farmer or rancher, first make sure your enterprise meets the criteria for being a centerpiece. If it doesn't, adding holons are not going to help you.

Rome wasn't built in a day and a high-income-producing farm or ranch won't be either. Start all new ideas small enough that if they fail it won't matter. A few of these may surprise you and grow into something far greater than you ever imagined.

The rules of the game:

Starting a holon

1. Protect the Mother Ship.
2. Seek sources of leverage.
3. Find the right person to be responsible for it.

Holon guidelines

- Holons should strengthen and reinforce the centerpiece enterprise and not compete with it for feed, labor or capital.
- A holon should primarily provide a service rather than additional income.
- A holon should scavenge labor and feed resources that would otherwise be wasted by the centerpiece enterprise.
- A holon should be "chunkable" in a crunch.
- A holon should require a minimum of new fixed overheads.
- A holon should require a minimum of extra marketing time to sell its production. The best customer for a holon's production is your own family.

Centerpiece guidelines

- It should be scalable.
- It should have inherent competitive advantages.
- It should produce cash flow on a regular basis.
- It should offer a high return to land and labor.
- It should be relatively weather-proof.

To learn profitability focus on the top 20 %

In enterprises where skill is required the top one to two percent often make 40% of the entire industry's profit.

One result of this is that the "average" producer who was once thought to be at the financial breakeven point is no longer at breakeven point but is actually losing money. In many industries only the top 20% are profitable at all.

Therefore, benchmarking yourself against the industry average is not only a waste of your time but is actually financially dangerous.

Let me put this into a sports context. If you want to learn how to win at baseball you don't study the teams in the middle of the league. Your time is better utilized by studying the play of the league winner and ignoring the rest.

This focus on the "average" seems to be particularly rampant in agriculture. A major problem this produces is that it makes the situation look far more hopeless than it actually is. Almost every farm or ranch can be made profitable if it is financially structured right but this correct structure is constantly changing and can only be discovered through benchmarking.

No doubt this middling focus serves those trying to make a case for increased farm subsidies but it is a major reason young people are no longer attracted to agriculture.

Young people aren't hearing that the top 20% of ranchers are producing a return on assets twice that of the average publicly held manufacturing business.

And that's with commodity priced animals!

How do the top ranchers do that?

A good way to find out is to read the annual ProfitProbe benchmarking study available from Ranch Management Consultants.

Ranch Management Consultants was founded by Stan Parsons and is now very ably being run by David Pratt of Fairfield, California.

ProfitProbe does a benchmark financial study of cooperating ranches in the USA and Canada. Thanks to a sister organization in Australia, ranchers in all three countries are able to compare practices and results.

Return on Assets is used as the benchmark. While this is not profit, it is a good judge of economic efficiency.

ROA is determined by dividing total income (cash and non-cash) by total closing assets. It can be increased by increasing income or by reducing assets.

The way to win this game is to produce a lot of cash flow from a small asset base. As a result, ranchers with leased land fare far better than those with deeded land and ranchers who custom graze or lease their cattle fare better than those who own their cattle.

Quoting Bud Williams, "Cattle ownership is a return to marketing skill. If you have only average marketing skills, you will make a lot more money custom grazing."

This survey confirms Bud's observation. The custom graziers fared far better than those who owned their own cattle.

While the ranchers who submit to such public financial analysis cannot be considered "average" because most of us would never do this, the same distribution curve of the top 20% of producers being way ahead of the "average" held true.

I am going to use 2002 results here because it is a more normal pre-BSE year. To further level the playing field, all livestock inventories were valued similarly, set rates of depreciation were used and unpaid or underpaid family labor was marked up to predetermined wage levels.

Here are the 2002 results.

The top 20% had a ROA of 11.7% versus only 1.4% for the average. However, it was actually worse than the average as half of the ranchers in the study had a negative ROA.

According to Pratt's analysis, the main thing that determined the level of return was how well the chosen enterprise matched up with the particular environment of the ranch. For example, cow-calf was a poor choice for short season, heavy snow areas. Such poor matches produced low ROAs because overhead costs rose to offset the inability to direct-graze.

However, ranchers who chose seasonal programs such as stockers in these environments were fully competitive with no-snow climate ranchers.

Productivity is directly related to the carrying capacity of the land. The highest productivity came from land capable of grazing one SAU (Stock Animal Unit - the equivalent of a 1000-pound dry cow) on three acres or less.

However, in this study these climate-blessed ranchers typically dissipated their natural advantage with higher equipment and labor costs per SAU. This allowed extensive ranchers (greater than 20 acres per SAU) with minimal machinery and high labor efficiencies to be fully competitive in ROA.

This tendency to bid away competitive advantage by over-investing in overhead is seen all over the world.

A personal survey by Jim Gerrish found that there was very little difference in the period of hay feeding between average cow-calf producers in Minnesota, Missouri and Mississippi even though the difference in the amount of snow cover was huge.

It was Jim's conclusion that it is the amount of hay made that determines the amount fed and not the climate. I think he is dead-on. I know ranchers in Florida who feed hay five months of the year in climates where it takes several decades to accumulate an inch of snow.

Interestingly, weaning weights and price received, which we tend to talk the most about had little to no effect on profitability. In fact the low ROA ranchers had an average weaning weight 27 pounds above the high ROA ranchers. Apparently, these higher weaning weights were purchased through higher feed costs as these producers spent considerably more on feed (hay) than the high ROA ranchers.

Also, seven out of 10 of the high ROA ranches received a lower than average price for their cattle. However, custom grazing with no cattle ownership on leased land had the highest ROA. Similarly, while scale is important, the highest ROA ranches were not the largest. Pratt said this seems to indicate there is an upper limit in the advantages of scale.

The overhead ratio is the main driver of net profitability as opposed to ROA. This is the ratio of things on your ranch you mean to keep versus things you mean to sell.

The breakpoint for profitability in this study was 50%. Every profitable ranch in the study had an overhead ratio of less than 50%.

Gross margin, that is the ratio of cash income versus direct cash costs, was not as large a profit driver as the overhead ratio but was best when kept to no more than 50%.

While most ranchers aspire to land ownership, this study found that as the percentage of deeded land increased, the ROA decreased. Ranchers who leased in excess of 76% of the land they grazed had an ROA that was 500% higher than those who grazed all or nearly all deeded land.

This leased land group had an ROA of 14%.

For ranchers with 100% deeded land, a gross cash flow that is equal to or better than 25% of the appraised value is considered exceptional. In other words, $1000 of land would produce $250 worth of gross beef sales per year.

While leased land ranchers have an annual rent to pay and deeded land ranchers don't, the renters still tended to be more profitable. Also, land lease payments didn't seem to have much effect on the gross margin ratio either. The bottom line was that owning land was a net negative.

And if you really want to shoot yourself in the foot financially, add a little crop farming. In the study, ranches with no farming had a higher ROA than businesses with farming. Yes, this includes the crop subsidy payments as well.

This low return is because of the high overheads incurred in maintaining seasonally used facilities and equipment. Interestingly, the average overhead ratio didn't change much as the proportion of farming in the mix changed. It is the first acre farmed that adds the biggest proportion of new overhead costs.

This facility and machinery cost called "plant" tended to decline in ranching as the ranchers' land quality and stocking rate increased.

In other words, ranchers in less favored environments tended to have higher infrastructure and overhead costs per unit of production than areas capable of carrying more cattle per unit of land area. However, the study found that the highest ROA producers in the low, medium and high groups were very similar. This indicates that good managers in low carrying capacity areas can compete with the humid areas by eliminating extraneous overhead costs.

In other words, management trumps location.

In fact, Pratt surmised that all of the low ROA ranches could become high ROA ranches by slashing overhead costs and creating an enterprise mix that fits their environment. In heavy snow country this may mean a shift to seasonal enterprises with custom grazing being one of the best options.

Once overheads have been lowered and gross margin has been improved, these ranchers can try to increase gross cash flow through the implementation of pasture subdivision or leasing additional land.

Note the recommended sequence:

Slash overheads, improve gross margins, and **then** seek to increase production. Adding new overheads to increase production just worsens the situation if you are already overcapitalized.

So how does the top American rancher and the top Australian ranchers compare? Pretty favorably.

In fact, the Americans are slightly more economically

efficient because they tend to utilize custom grazing more, which increases asset: turnover ratio.

However, the best American ranchers still tend to have more overhead "stuff" than the best in Australia or Canada, which have similar ratios.

Despite the large numbers of animals per man we hear about in Australia, the labor efficiency benchmark between the USA and Australia were very similar and gross margin ratios were virtually identical.

Canada's gross margins were less because their best ranchers tried to hang onto their stock during the 2002 drought and fed more hay whereas the Americans and Aussies tended to destock earlier and avoided feeding very expensive hay.

Pratt said that many ranchers don't like the survey results because they do not include land appreciation which would considerably sweeten the returns. However, Pratt said many ranchers in the survey ranching on rapidly appreciating land with no land cost were still losing money on their cattle on an out-of-pocket basis. He said even if you are getting rich on your real estate investment, it seemed to him you would want to staunch such a financial drain.

In conclusion, Pratt said the annual survey continues to show that the keys to economic efficiency are to get your enterprise in sync with Nature and to relentlessly attack overhead costs. This formula will work on almost any ranch, anywhere.

The rules of the game:

1. One of the best options in heavy snow country is to shift to seasonal enterprises such as custom grazing.
2. After lowering overheads and improving gross margins, add more pasture subdivision or lease additional land.

Surviving the shake-out years

The primary problem for all producers in mature industries is surviving those oversupply, shakeout years. Things cannot get better until someone quits and there are few volunteers for bankruptcy. Because of the efficiency of the capitalist system in routing capital to profits, most producers will spend the majority of their lives in the oversupply mode.

The widely touted solution to this problem is a contra-cyclical diversification that will be up when your main thing is down. For example, economic research in New Zealand has found that a 50/50 (by body-weight) split between sheep and cattle does indeed stabilize returns at an almost constant level of six to eight percent year in and year out. This return is comparable to American blue chip corporations.

In Australia, a similar income stabilization is achieved by wheat farmers who are also into sheep. Wheat and sheep numbers expand and contract down there in response to whichever enterprise offers the best return.

In Argentina, grain farming and the grass finishing of beef cattle are the two contra-cyclical activities of commercial scale farmers in the Pampas.

It is interesting to note that all three of these proven diversifications are from the Southern Hemisphere where the government does not provide farm income stabilization.

Not so many years ago, grain farmers in North America achieved similar price stabilization by being diversified into both grain farming and market hog production.

When corn was cheap you walked it to town on the hog's back. When corn was high the sows went to slaughter and the grain to town.

This proven diversification fell victim to the government's set aside and subsidy programs that encouraged specialization in pure grain production. This production simplification has subsequently proven to be hard on the soil, the farmer and the taxpayer.

Let's take a look at how we get ourselves into such traps.

First, diversification requires that not all of your land can be utilized for whatever is the most profitable use for it at that particular moment in time. This sets us up for attack from "efficiency experts."

Flexibility is always the opposite end of the teeter-totter from efficiency. You can't be both totally flexible and totally efficient.

Flexibility requires an alternative enterprise be kept operating even when it would be more profitable to just chunk it and put all of your capital into whatever was currently most profitable in the short-term. Falling production margins caused by the "efficiency" of production simplification helped make farmers increasingly susceptible to "short-termitis."

Second, production simplification suited the needs of the agri-input industry far better than diversified farming. That continuous cropping worsened weeds and pests was all to their benefit. Today, 90 cents out of every dollar received for commodity grain goes to the input suppliers, shippers and handlers.

Third, most university extension and research personnel are now "specialists" and very narrowly trained. Consequently, almost all extension information is presented in the context of highly simplified mono-enterprise, production systems.

These high-cost, brittle production systems only work when prices are high. Unless we have a major drought or war, commodity prices are more likely to be low. In that scenario, only the least-cost producer survives.

But we are not interested in just survival. In North America, we and our children have all sorts of other options. Agriculture has to be competitive in the income it provides or we won't do it.

It may be that clean water and cleaner air will be our primary product in the developed world. If you study the Argentine production system, it is precisely what is currently being proposed as a way to lower atmospheric carbon dioxide.

Let's take a look at it.

The key way in which the Argentine system is superior to the old Midwest corn and feedlot system is that the grain crop can be — and usually is — harvested directly by the cattle.

> *A successful sustainable business lives by its principles. It is prosperous without being wasteful. It grows without mortgaging its future. It shares its discoveries without giving up its leadership.*
>
> *Paul Dolan*
> *True to Our Roots*

In contrast, the Midwestern system requires that the grain be mechanically harvested, dried, stored, mechanically fed and the manure mechanically gathered and spread. Note how much more energy (money) the Midwestern system requires.

To give you some idea of the capital lowering consequences of direct harvest, the 30,000 acre Nueva Castilla Ranch near Santa Rosa that I visited has only six tractors, three planters and three grain drills.

The ranch plants all of their crops using their own machinery but the grain harvesting is done exclusively by contractors. This lack of investment in harvesting machinery helps keep the direct-graze option attractive to Argentine ranchers.

Nueva Castilla grows corn, wheat, rye, sunflowers and soybeans plus enough pasture to grass finish 23,173 steers each year. The ranch also has a "scavenger" cowherd of 3000 head that exist solely to recycle crop residues, and clean up roadsides, swampy areas and conservation plantings. An additional 10,000 mother cows extensively graze the dry brushlands to the west of the grain/beef finishing ranch.

The owner of Nueva Castilla, Eduardo Pereda, said the diversified crop and cattle mix provided excellent year-round cash flow, work for his labor force and better equipment utilization. All planting is done no-til as his ranch is very susceptible to wind erosion.

He said they planted 21,600 acres of grain and pasture crops in the spring and 10,800 in the fall. The mixture of crops and forages allow a few tractors to cover a lot of ground.

He said the decision on whether to machine harvest or graze out a crop was made depending upon the price of grain. However, he preferred to direct-graze his corn and small grain crops.

All crop land is rotated to permanent alfalfa/grass pasture for 4.5 years out of each ten to rebuild soil organic matter and nitrogen. He said his sandy soils were naturally very low in organic matter with most of them measuring less than two percent.

The cropping/finishing ranch is divided into 110 permanent paddocks with one-wire high tensile fence. New Zealand-made fence energizers are used for power. All vehicles are provided with front-mounted skids to allow them to drive over the fences. Gates are provided anywhere in the fence by insulated poles which hold up the wire.

The cattle are rotated on a three- to seven-day rotation with the exception of the spring when they are set-stocked. The Argentines believe set-stocking helps keep spring pasture under control better than a fast rotation.

Some corn is made into silage and some spring pasture is harvested as pasture silage. These silages are fed underneath the one wire electric fences in the winter to keep steer gains from falling below one pound a day. Pereda said he felt that allowing finishing animals to grow too slowly in the winter resulted in big-frame animals that were then difficult to finish.

He said the secret to making money with grassfed beef was to maximize the utilization of the spring lush. Consequently, he buys and ships in around 10,000 stocker calves from the subtropical north of Argentina in addition to his home raised calves to get his spring stocking rate higher.

Since his main cow-calf ranch is in the semi-arid West his predominant breed is Hereford. He sells his finished steers on the domestic market at around 1100 lbs liveweight. He likes to ship some kind of cattle (finished steers, cull cows) to market each week for cash flow.

He said he personally preferred to graze corn rather than machine harvest it. He said grazed green corn was used to finish steers during the late summer slump in pasture production.

Here are a few lessons I think we can learn from Argentina:

1. Rotating land through a relatively lengthy period of grass/legume permanent pasture is an excellent way to get grain crop yields up and pest and weed control costs down. Virtually every major weed of crops is an excellent forage and virtually every major pasture weed is controllable with tillage.

2. The forage finishing of beef cattle on high quality land and annual crops is economically competitive with grain production.

3. The best way to bring grain prices up is not to harvest it as grain and feed it to ruminants in feedlots but to take the ruminants to the field and let them harvest it in its green pre-grain stage. We have long known this with wheat and oats, but it also applies to corn and grain sorghum.

4. No-til systems are better than tillage with direct grazing in that they keep the soil firm and prevent winter pugging. No-til does not necessarily require herbicides. The Argentines will use dense shade-producing forage crops such as sudangrass to remove difficult warm-season weeds.

5. Cereal rye can provide green winter forage in even relatively cold climates. Most summer cropland can also grow a winter forage crop.

6. Direct grazing allows for the use of grain species such as oats that are useful in rotations but for which there may not be a local cash grain market.

7. The spring lush is your primary profit generator. Don't waste it by being understocked.

8. The Argentine system seems ideal as a production model for a high value organic grain/grassfed beef program.

Thinking the unthinkable

A well-read book of mine is *The Art of Profitability* by Adrian Slywotzky.

I typically re-read it every few weeks or so as it is very information dense. However, that doesn't mean I necessarily like its core message.

This message is that every business, industry and economic activity starts at a level of zero profitability for everyone involved, traverses an arc to an apex of peak profitability for everyone involved and eventually returns to a level of zero profitability for everyone involved.

I found this very sobering.

Most of us prefer to think in terms of an ever upward trending chart line in which the good times only get better. That we will eventually have to periodically totally recreate our businesses just to survive is daunting.

The Spanish have a word for the art of living life to its fullest while remaining fully aware of one's eventual death. This word is *duende*.

And, I think that is what all of us have to learn to do.

We have to enjoy the good times but realize that they will eventually end.

I was reminded of Slywotzky's arc when I got a press release on Peter Apedaille's keynote speech to the Christian Farmers Federation in Canada.

Apedaille is professor emeritus in the Department of Rural Economy at the University of Alberta and a founding member of the Canadian Rural Revitalization Foundation.

He said that after 50 years of decline, net farm income in Canada is now approaching zero. This is due to a combination of a high Canadian dollar and farm overcapitalization.

Things are not much better south of the border.

The *Wall Street Journal* pointed out that despite record yields and government subsidies of $30 to $50 an acre, the average acre of corn in Nebraska lost $88 an acre in 2005.

A key point Apedaille made is that there is little the government can do to "fix" what ails North American agriculture.

For example, what happens to the billions of dollars the taxpayers pour into agricultural subsides each year? The majority of them are bid into inputs, the biggest of which is the price of land.

$1000 an acre of the price of land in Iowa is said to be due to capitalized government corn subsidies. Soon, it will be $2000 an acre. Then $10,000 an acre.

The longer subsidization goes on the more unrealistic the price of inputs becomes and the less competitive you are as return on investment shrinks.

Government subsidies are like coal to a boiler. If you just keep stoking, you will eventually blow the whole thing up.

The last time this happened was in the early 1980s. This explosion released the pent-up pressure of over-capitalization and allowed the process to slowly start over again.

25 years later we found ourselves back at the same level of over-capitalization, and the government is stoking away, thinking that more coal is what is needed to keep the darn thing moving.

Remember the old saw, "The unintended effect of government action is always far larger than the intended affect."

Apedaille said the best government agricultural policy could do was to try to manage the decline. It cannot stop it.

The forces of globalization are ensuring this is not just another cycle but probably the end game for undifferentiated commodity agriculture in North America.

To have a sustainable profit cycle you must have three things:

1. An exceptional product.
2. A profitable business model.
3. A source of control over your competition.

Why are doctors, lawyers, dentists and pharmacists highly paid?

Because they control their competition with rigorous licensing requirements.

The vast majority of what you pay a doctor for could be done just as well by a well-trained nurse or medic. But, just try to practice medicine without a license!

If you want to sell your production wholesale for the highest price you need to be Certified Organic. Why?

Because the hassles of organic certification offer an effective competition control!

You don't think non-organic Nebraska corn farmers who sold their corn for $1.60 a bushel in 2005 wouldn't have preferred to have sold it for the organic price of $8.00 a bushel? The certification program kept the supply in 2005 restricted despite a massive oversupply of corn.

Colorado sheep rancher, Richard Parry, said that he went to a Certified Organic program for the competition protection it provided but also added an Argentine-style year-around forage chain sequence to eliminate having to buy expensive organic alfalfa hay.

That's the way to do it!

Use the certification program to build a moat against competition but keep your production input costs low by using Nature as a guide.

Custom grazing with livestock

Custom grazing is more profitable for the average landowner than owning cattle.

A 2002 benchmark survey by Dave Pratt of Ranch Management Consultants showed that custom graziers consistently had the highest return on assets of the ranchers studied with a 14% return.

This survey also found that ranchers in climatically disadvantaged areas could be financially competitive if they concentrated their efforts on the season when their grass grew best - or at all.

Custom grazing is the best way to do this.

While we tend to think of custom grazing in terms of stocker cattle, it also works well with cow-calf. The average cow-calf producer would be far better off paying a professional grazier $300 a year to graze and manage his cows than to do it himself.

Conversely, receiving a check for $25 per cow each month would be far more attractive to many ranchers than trying to ration out one big check a year.

The one big advantage of custom grazing I have seen is that when ranchers are freed of the supposed necessity for out-guessing the market, they can concentrate fully on the grass and grazing management.

They are no longer in the beef business. They are in the grass business and they tend to get really good at it.

While this all seems to make good economic sense, I must tell you that this industry has generally not fulfilled its promise for either the grazier or the cattle owner.

Graziers get cattle that have been fed grain and won't gain well for six weeks until they get their rumens straightened out. Quite often with spring delivered cattle this results in the grazier missing his best gains of the season.

Conversely, cattle owners occasionally get back steers after months of grazing that weigh less than when they delivered them. This is particularly true when perennial forages are used.

Graziers are afraid they won't be able to find clients, and cattle owners are afraid they won't be able to find competent graziers.

Rates per head and per pound roller-coaster from year-to-year based upon the price of grain and drought. This pricing volatility makes it hard for either party to make long-term plans.

All in all, it is an industry fraught with anxiety.

For all of its potential, custom grazing will not go anywhere fast until we can unravel the knots in the stomachs of both the grazier and the cattle owner.

The best way to do this is to increase the information flow between the two.

The word fear is said to be an acronym for "False Expectations Appearing Real."

Grassfed beef specialist, Anibal Pordomingo, said the primary tool of the professional grazier is a set of livestock scales. If your business is selling gain, you need to know how much they are gaining. You cannot improve what you cannot measure.

In the majority of the cases, poor animal gains at the end of a season are as big a surprise to the grazier as they are to the livestock owner.

It is not necessary to weigh every animal every month but a representative sample needs to be weighed.

These monthly weigh figures would not only reassure the grazier, they would also reassure the livestock owner. These weigh figures can also form a basis upon which a monthly payment can be based.

With most perennial grasses, up to 80% of the total gain is produced in the first 20% of the season. Once daytime temperatures consistently get above 86 degrees F, gains on perennial grasses tend to collapse.

The New Zealand system of selling one-third of the cattle in (the Northern Hemisphere equivalent) June, one-third in July and one-third in early August is a good one and works equally well on both cool-season and warm-season perennials.

> *You don't earn a living. You learn a living.*
>
> *Jim Botkin & Stan Davis*
> *The Elephant Under the Bed*

Statistically, selling feeder cattle in the middle of the green season is almost always more profitable than selling them at the end. This makes perfect sense because so few of us currently do this.

Of course, famous Mississippi grazier, Gordon Hazard, has sold all of his cattle in early August for many years and allowed his fescue-dominant pastures to stockpile growth. This has allowed him to develop a low-cost, hay-less, production system.

Missouri custom grazier, Greg Judy, also follows this seasonal pattern.

This system of a summer sale and stockpiling the pasture also allows the client to buy calves in the fall when they are cheaper and have you over-winter them on stockpiled forage.

Late summer fescue is worth a lot more producing a pound a day in January than a half pound a day in August and the rates charged should reflect this.

Another benefit of this production cycle is that the grazier gets to take a four to six week late summer vacation each year. August is not a bad month to miss where I live.

While many ranchers would like to graze livestock for

other people as a way to avoid capital costs and market risks, few consider the benefits of taking the other side of the deal.

Buying custom grazing allows the livestock owner to buy both land and labor on a per head basis.

In other words, the custom grazing client is always in a perfectly balanced production "sweet spot."

Also, when the market hits the fan the buyer of custom grazing services can easily bail out because all of his costs stop the day he sells the cattle. In contrast, many of a landowner's costs continue whether there are cattle on the property or not.

A very viable alternative to obtaining more land is to start putting your cattle out with others on a custom basis.

Equally viable is a grazier whose land and climate are best suited for cow-calf, buying stocker and grass finishing gain from graziers in more favorable climates.

Terry Gompert, who works a full-time day job in Center, Nebraska, said the short day-length days in winter made it impossible for him to see his 100 cows either before or after work during the week.

"There's just no sunlight when I get home from work," he said.

"If it weren't for custom grazing, I'm not sure I could be in the cattle business."

Gompert said paying a corn growing neighbor to graze his cows on stalks is a win-win situation for both.

He said figuring both stalks and transportation, he is paying 48 cents a day to have his cows custom wintered from October 15 to March 15.

This compares with $1.43 per day if he bought hay and fed it at home and 65 cents if he rented stalks and grazed them himself.

"Most people don't charge for their own labor. If you did, you'd find custom grazing is always cheaper."

He said the farmer wintering his cows sees custom wintering as a good deal as well.

"He is wintering his own cows on the stalks, so there is

very little more effort to take care of my cows as well.

"Plus, research has shown that grazing the stalks increases subsequent yield by seven percent."

An extra benefit of winter custom grazing is that it lowers the taxes each cow has to amortize because less land is needed per cow.

"In Nebraska, land taxes are 50% of the pasture cost of the typical producer."

Gompert said extension research has shown that land taxes range from $20 to $100 a cow in Nebraska.

This variation is primarily dependent upon stocking rate with irrigated pasture having the lowest taxes per cow and rangeland the highest.

His land taxes on his non-irrigated ranch are $70 a cow but would be nearer the upper range if he had to winter the cows on forage made from, or stockpiled on, his own land.

He said there is currently a $300 a cow range in the cost of production between the high and low cost producers in Nebraska.

"You've got to look for the biggest bang for your time and money. The biggest cost saver is to eliminate having to make and feed hay."

He said that corn stalks were always cheaper than hay and particularly so in drought years.

"Switching your calving season from February to May can cut your hay feeding costs by $100 a cow."

His management goal is to spend no more than $208 a year to produce a calf. He said such a production cost is competitive with the best cow-calf producers anywhere in the country.

"Our great unfair production advantage in Nebraska is an abundance of cheap corn stalks. We make our money here in the winter, not in the summer."

This is particularly true of brood stock where grazier skill is less important than stockers or finishing.

Another way to lower land costs is to "stack" non-competitive species rather than to expand with just one.

With a cattle based operation, adding sheep to utilize broadleaf weeds or goats to utilize browse are very profitable as long as their stocking rate matches their particular forage niche.

New Zealand economic research found that a ranch balanced by bodyweight with equal parts of beef cattle and sheep showed a constant six to eight percent return.

> *What gets measured gets done. What gets paid for gets done more. What gets paid well gets done more...and better.*
> *Tom Peters*

However, stackable enterprises do not have to just be ruminants.

Joel Salatin keeps the output of his home farm high with pastured pork and poultry enterprises that heavily utilize off-farm grain purchases.

An alternative is the "holonic" approach whereby alternative species live off the waste generated by the center-piece primary enterprise.

A good example of this are whey-fed pork and poultry living off the free byproduct of on-farm cheese production.

Another is free-ranging poultry, which live largely on insects, and which are attracted to the grazing animals' manure.

To keep labor needs down, you want to concentrate on animals that can largely gather their own feed or self-feed from a large stored feed source.

Hunting rights sales are another profit improver and wide-spaced timber production that integrates grazing and timber another. The possibilities go on and on.

Once you have reached the limits of your creativity of generating income with your existing property, you can start to consider a further expansion.

The whole idea is to first concentrate on growing revenue without adding new fixed costs.

If an enterprise requires much new capital investment, it should be looked at very carefully.

However, you cannot maintain or expand your income without growing in volume. Anything that can be replicated at a profit, will be, and will fall in price over time.

The rules of the game:

1. The biggest advantage of custom grazing is freedom from out-guessing the market.
2. Custom graziers are in the grass business, and they tend to get really good at it.
3. It is not necessary to weigh every animal every month, but a representative sample needs to be weighed.
4. Buying custom grazing allows the livestock owner to buy both land and labor on a per head basis.
5. One way to lower costs is to stack non-competitive species rather than to expand with just one.
6. First concentrate on growing revenue without adding new fixed costs.

A custom grazing success story

Missouri custom grazier, Greg Judy, described his custom grazing operation as "management-intensive" even though both he and his wife have 40 hour a week jobs in town.

Judy recommended that new custom graziers not start with stocker cattle but dry cows.

"If you do a good job with a man's cows, stockers may follow."

He still grazes some cows because when he leases a new farm the pasture is not of stocker quality.

"You don't want to be grazing stocker cattle on fescue without legumes. Most of the types of farms you can lease cheaply will require liming, fertilizing and subdivision before they will grow acceptable legumes."

He started off charging 28 cents for a pound of gain but now charges 32 cents in the green season and 32 cents per pound of gain plus 18 cents a day yardage for the January 1 to March 1 winter season.

"Wintering cattle is a lot more work and you should get paid for it."

He winters stocker cattle on strip-grazed stockpiled fescue plus 1/4 to ½ of a large roundbale of hay. All hay used is purchased.

He said the stockers are managed during this mid-winter period so as to gain around ½ pound per day. This results in a high compensatory gain in the spring.

"Wintering cattle gives you cash flow during those months plus allows you to get your cattle properly conditioned for the spring greenup. When the grass comes in March you've got yourself one supreme mowing machine. A real tornado!"

Judy said he liked the seasonal flexibility of stocker cattle but warned against getting fixated on just one class of cattle. When calves are expensive stocker clients will be hard to find but cows will be easy.

He watches the news for drought areas and advertises his services there.

Cows are charged $15 a month in winter, which he said was a bargain compared to wintering the cows on hay. Cows are charged $10 to $15 per month in the green season. Unlike stockers, he will take cows through the August summer miasma.

Bred heifers are $10 to $12 a month and Judy really likes this class of animal because of its minimal health problems.

He has grazed horses for $20 a month but warns that they are very hard on pastures and need more than minimal one-wire electric fences.

"You need to use white polytape if you are going to graze horses as they find it hard to see a metal wire."

He said horses are useful to help clean up a new farm as they will forage on brush and saplings similar to a goat.

He said it was absolutely imperative that you have a written contract that covers all contingencies including death loss. When he started out he took no death loss liability, but now that he is confident of his skills he splits any death loss over three percent with his clients. He said he has never had to pay this yet.

He also requires that all clients carry liability insurance in case the cattle get out on the highway and cause an accident or gore a trespasser.

Judy said clients are billed for half of the gain the cattle make each month and the remainder is paid when the cattle are picked up. He tries to arrange the trucking so that a truck bringing calves in carries yearlings out. He has found he can get his clients a better rate on trucking this way.

"Keep in mind you are in a service business. I try to call all of my clients once a month and let them know how their cattle are doing. Most of them never come to look at their cattle until they are ready to be shipped."

Some services he provides for clients are:

Sorting off of heavy calves for sale and weaning, worming and wintering calves for cow-calf clients.

"If you'll lease ground and custom graze cattle you can be in the cattle business in a big way without having any money. There's lots of idle ground that people own but don't know what to do with. They don't have the energy to learn MiG themselves and want someone else to do it for them.

"If you will develop your reputation as a hard working, honest grass farmer, both cattle and land will come to you. You won't have to go looking for them."

Judy said his current plans are to build up his custom herd to around 1500 head. At that level he plans to quit his day job and become a full-time rancher.

"I make about $40 an hour looking after other people's cattle for them. I have no risk of cattle prices dropping. No risk of death loss. My $40 an hour is a pure return to management. Custom grazing is not a bad job."

Advantages of custom grazing

- Monthly cash flow.
- No equity required.
- No interest payments on cattle.
- Can load up pastures in spring to keep seedheads off.
- Tremendous short-term flexibility to match rainfall variability.
- Can completely destock in August to build fall grass stockpile.
- Very little hay needed.
- No mineral, death loss, trucking, medicine or insurance costs.

- Can grow as fast as your management skills allow.
- No risk from falling cattle prices.

The grazing contract should cover:

- Date that stock will be delivered and out date.
- Weigh tickets at arrival and departure.
- Death loss responsibility.
- Grazing charge and payment schedule.
- Who pays for mineral, medicine and trucking.
- Grazier's responsibility for doctoring sick cattle.
- Liability policy on cattle.
- Cattle owner and grazier signatures with date of signing.

Livestock ownership not necessary for successful ranch

Omaha investor, Dr. Lon "Mickey" Keim, said his 12,376-acre ranch was bought and run as a real estate investment more than a livestock business.

In the late 1980s he became interested in buying a ranch to graze a herd of cows he had put together over the years. He saw the tough times of the early 1980s knock Sand Hills land prices to less than $100 an acre. However, by 1989 they had recovered to $150 an acre and he thought they were clearly on their way up. Consequently, he decided to buy the 7241-acre Brush Creek Ranch (enlarged to 7416 in 1991) in the Sand Hills region near the South Dakota border.

He said that ranch investors like himself make their money in two ways — income from operations and income from land appreciation. However, to an investor with a sizable off-ranch income, land appreciation is more important because it is tax-free until sold.

"As an investor, you buy a ranch primarily because you think it has appreciation potential, not necessarily because of its cash flow. I don't need the ranch's income to live on."

However, he said he did not ignore increasing the ranch's potential for increased cash flow from operations either.

Because of his row crop background, he initially focused his attention on maximizing the return from the two quarter-section center pivot systems already in place on the ranch. He put money into grain handling facilities and bin storage - investments that he quickly regretted.

"I realized that I wanted to grow a dependable, reproducible crop every year without the recurrent input costs. Focusing my attention solely on pasture and range management was the answer."

As a result, he now leases out the two pivots to neighboring corn farmers while retaining the grazing rights to the stalks in the winter. The rent on these two pivots pay the taxes on the entire home ranch. ($26,335.70)

He realized he had neither the background, expertise, nor the time to learn the nuances of livestock marketing, breeding and genetics. Consequently, he decided to specialize in custom grazing cattle for livestock owners who needed summer range.

"We are in the business of selling quality pasture management" is how he defines his ranch's chosen niche.

Nebraska is perhaps the heaven of custom grazing in the United States.

Many corn farmers in the state own cow herds to utilize their corn stalks in winter but do not have any personal summer grazing land. They are too busy with crop production during the growing season to take care of their cows, and seek out graziers willing to provide both grass and care for the corn growing and harvesting season.

In 1989, Keim started with 650 cow-calf pairs charged out at $75 for the five-month-long green season. The ranch initially only had 13 paddock subdivisions and was largely dependent upon live water sources such as streams and natural ponds for stockwater. He estimated that at least 80% of the ranch's surface was bare ground at that time.

Attending a Ranching for Profit School, he met Texas rancher and consultant Bob Steger. He subsequently hired Steger to create a comprehensive development plan for the ranch and to oversee its installation and management.

Each year Steger draws up a grazing plan for the ranch and sets the ranch's stocking rate depending upon the grass and soil moisture conditions.

Steger's ranch development plan started by putting in a central two-inch stockwater line down the center of the ranch. From this central line, stock tanks could be plumbed in as additional paddock subdivisions were created. Taking advantage of natural folds in the land, more ponds were bulldozed into being as well.

The ranch has a four-strand barbed wire perimeter fence with a "shadow" electric wire on outriggers. The interior fencing is all two-wire aluminum. Both solar and 110-powered fence energizers are used.

Keim said a major initial mistake was the use of pie-shaped paddocks radiating out from the water points. These pie-shaped paddocks created very poor grazing distribution. They have subsequently been torn out and replaced with square or nearly square paddocks with the stockwater access in one corner. Currently, there is one stockwater tank for every five permanent paddocks.

A 36-foot corridor around the stocktank was found to be the optimum width to allow cattle to water without crowding but discourage them from loafing at the water point. Metal stocktanks with concrete bottoms 38 feet in diameter are preferred. Tanks of this size have been able to water herds as large as 1500 head with no problems.

Removable metal panels prevent cattle from getting into the tank but allow for easy tank cleaning and maintenance.

Over the years the paddock subdivision has been gradually increased by making the paddocks ever smaller. There are currently 108 permanent paddocks. Keim now considers a paddock 40 to 60 acres to be nearly ideal in size.

A paddock this size is typically grazed with a herd of 450 pairs for one to two days. This gives a high degree of stock density to which the native Sand Hills warm-season grasses respond exceptionally well and which actually increases grass growth.

An early first-hand lesson was learned when he placed 1565 yearlings on a 26-acre sub-irrigated hay meadow for one day. That unfertilized meadow subsequently grew far more hay after having been "mistreated" in this manner than equivalent acreage under a heavily fertilized center pivot.

"This land was meant to be grazed at very high stock densities for very short periods of time. We have been very gratified to see the warm-season perennials returning with the proper grazing management. We've learned that if it ever grew on your ranch the seeds are still there and they will come back."

One warm-season species that has popped up and appears to be doing well is dallisgrass. Since this grass is traditionally seen in the blackland prairies of Texas, Mississippi and Alabama, its ability to grow so far north in latitude was previously unknown.

Keim said the grass cover on his largely, sand dune ranch, is now estimated to be 70 to 80 percent and is a mixture of both cool-season and warm-season plants.

An attempt to increase stock density by increasing herd size created a lot of stress problems when the herd was over 600 head. There have not been any problems at the 450 head herd size.

In 2000, thanks to its improvements, the ranch was reappraised at $315 an acre. This meant that the ranch's value had more than doubled in 10 years.

Keim "harvested" his accumulated capital gains via a new mortgage and spent them on an additional 4800 acres. Since this over-grazed and unsubdivided land was appraised at nearly $100 an acre less than his home ranch, Keim believed that the two new ranches could be brought up to the home ranch's value with water and fence developments.

In other words, he thought it was a reasonable land investment from a probable future appreciation standpoint.

Fencing and water developments are not only deductible to current landowners but raise the tax sheltering depreciation basis available to a subsequent buyer as well.

This development basis is important as the land itself is not depreciable.

When a ranch is sold the depreciation schedule on the improvements starts over and they can be depreciated again. This makes the ranch a much better buy to a subsequent investor from a tax standpoint.

Therefore, to an investor, fence and water improvements pay in three ways - increased current income from a higher stocking rate, increased tax shelter from current depreciation and a higher depreciable tax basis to a subsequent owner, which should raise the sale price of the land and the subsequent capital gains.

Because the new land was adjacent to the current property, stockwater could be brought in by just extending the home ranch's main line across the boundary fence. Keim said he is definitely not interested in buying little dribs and drabs of land far from his home ranch.

In 2003, the home ranch's appraised value had risen to $360 an acre and Keim again harvested this by refinancing. Harvesting accumulated capital gains by borrowing it out of the property is much more tax efficient than selling and re-buying properties.

Thanks to low interest rates, the cash flow saved by the new mortgage was enough to pay the ranch's entire annual labor bill.

The ranch grazes 1700 cow-calf pairs, and charges $150 per pair for a five-month graze. This gives the ranch an operational income of $255,000 or roughly $20 an acre.

The ranch is run by its manager, Daryl Butterfield, and his two children, with seasonal help from local high school students. Butterfield is given an annual labor budget allotment. Any monies he does not spend on outside labor accrue to him.

Butterfield also has a calving out service separate from the ranch and charges $25 for every live calf delivered for the ranch's cattle owners. He is also allowed to run 35 cows on the ranch for no charge.

Keim said he has learned that the best way to reduce labor costs is to pay a few good people very well.

There are two 400-head herds and two 450-head herds. Each herd has its own set of paddocks and are shifted on a one-to-two day move.

The ranch sells its grass in 400-cow minimums. Next, he will go to 600 cows as a minimum.

"The fewer clients you have to work with the better it is," he said.

When he first started, cattle owners were leery of subdivided ranches as many pioneer MiG ranchers left them in a paddock too long and animal performance was often low.

"We now have a waiting list. I consider that our greatest compliment."

Keim said that by moving cattle every day the grass quality, rumen function and animal performance tends to be constant. If cattle are kept in a paddock for several days, they eat the best of the grass the first day, the next best the next day and so on. This variable diet creates a roller-coaster for the animal's rumen and subsequent animal performance.

Ideally he would like to graze the whole ranch with yearlings as that would maximize seasonal stocking rate flexibility and thereby stock density per acre. However, the ranch has had poor luck custom grazing yearlings.

He said today's yearlings tend to come in too heavy in the spring to be profitable for a grazier. This is particularly true of heifers which gain poorly when they weigh over 800 pounds.

Then there is the difference in rates. Even though an 800-pound steer is nearly a cow's grass eating equivalent, stocker owners will only pay 60 cents a day whereas cow owners will pay a dollar.

Another problem is that most stocker cattle in Nebraska are assembled in feedlots and are fed grain prior to being shipped to the grazier. This grain feeding destroys the rumen bugs necessary to digest grass and makes for low to no gains for the first month they are on grass - a month when grass quality is capable of three pounds a day gains.

Keim hasn't given up on yearlings entirely yet. He is

experimenting with a partnership arrangement with an adjacent feedlot.

Under this arrangement, the cattle will be assembled and backgrounded on a no-grain diet, moved to Keim's grass and then walked from his pasture into the feedlot for finishing the following winter. Any profits above the ranch's grazing fee and the feedlot's yardage and markup will be split between Keim, his manager and the feedlot owner.

"I dearly hope this works as I would love to have the whole home ranch grazed with yearlings rather than cows."

Custom grazing offers consistent profitability

Pete Ferrell said his real love was cow-calf production but after being battered by several cattle cycles he found custom grazing to be much more consistently profitable.

"I have never lost a dime custom grazing. I can't say that about my other livestock ventures," he said.

4L Grazing is known as "a management company that leases properties in the Flint Hills for livestock grazing."

He said the secret to success in custom grazing is "loyal clients and glowing referrals."

The Flint Hills area of southeastern Kansas is a tall-grass, warm-season prairie that, thanks to a limestone layer just beneath the soil, has never been plowed.

It is famous for the excellent gains on yearling cattle it can produce until mid-summer and has been a major custom grazing area since the late 19th century.

However, in recent years 4L Grazing has specialized in custom cow-calf grazing. Most of these cow-calf clients are farmers who use the cows as a way to harvest their crop stover in the winter.

4L Grazing offers both a 90 day May to August stocker program and a full season May 1 to October 31 cow-calf program.

"Cows are more profitable than stockers due to the longer season but the stocker program allows us to lower our overall stocking rate in late July when the grass growth dramatically slows down.

"The stocker areas are then allowed to regrow and the

grass is sometimes used with the cows in the fall and winter if we have adequate residual," he said.

4L Grazing offers both "brown season" and "green season" grazing for both stockers and cows with the brown season priced at a considerable discount to the green season.

"Any brown season grazing fees we earn are just gravy."

Each "green season" stocker contract is sold on the basis of tonnage of incoming weight, for example, 198,000 pounds plus or minus two percent might satisfy one contract allotment. Each allotment is based upon the normal carrying capacity of a particular parcel of the ranch.

4L Grazing charges a 90 day rate of $13.50 per incoming cwt for their stocker allotments.

The calculation for the 90 day rate for a 450 pound steer would be 4.5 X 13.50 or $60.75 for the 90 day graze.

Ferrell said it is the responsibility of the stocker owner to provide the gross contracted weight to fill an allotment. If he provides less than 198,000 pounds, the client is still charged the full contract tonnage.

"In this way I don't care what the head count or the weights of the individual steers are. We weigh them coming in and we weigh them going out."

Research in the 1970s found that 80% of the total season's gain occurred in the first 90 days and if destocked in late July, twice as many stockers could be carried on the same land area.

This has subsequently become the norm for stockers in the Flint Hills.

A $10 per head non-refundable advance against future grazing fees is made when the grazing contract is signed.

Ferrell said both the demand for grass, and the fee paid, for stocker grazing is highly dependent upon the cost of gain in the feedlot, which is dependent upon the price of corn.

Cows and bulls are charged $115 for the green season and the calf is charged $50.

"We don't take in any cows younger than four years of age and we want all the cows on our place prior to calving."

Each year the stockers and the cows switch grazing areas on the ranch and the direction of rotation to prevent hitting the grass the same way at the same time of the year.

The stocker cattle are set-stocked for their 90 day graze but the cows are rotationally grazed.

To facilitate this the ranch uses a lot of "seasonal fencing" (temporary electric) in addition to permanent paddocks to vary the grazing patters for the same reason. This fence is all taken down in the winter to facilitate pasture burning.

A three-year burning rotation is used to prevent cedar encroachment with one-third of the ranch fired each year.

Until recently, 4L Grazing leased an additional 5000 acres and plans to do so again. Ferrell said the primary problem with most leased properties was the lack of stockwater development.

With both the cows and stockers, mineral supplements and medicines are charged to the client at no markup.

"Some people try to make the selling of medicine a separate profit center but we don't. We want them to know that we make our money solely from grass management."

Brown season grazing is $10 per adult per month.

He said in years past there was little demand for brown season grazing but the high cost of fuel has increased the demand.

"At $3.30 a loaded mile, they no longer want to take the dry cows home to corn stalks. They want to leave them here."

Custom grass finishing — The ultimate skill

Allen Williams, Chief Operating Officer of Tallgrass Beef, a major grassfed brand, said the major problem restricting the growth of the grassfed industry is not a lack of customers for their beef but a lack of graziers who know how to finish cattle on pasture.

"Professional grass finishers are extremely hard to find. We have found that cow-calf and stocker expertise just does not translate into finishing expertise," he said.

"It's an entirely different mindset."

While Tallgrass does offer separate contracts for stocker growing and finishing, they prefer to use graziers who can perform both functions.

He gave the following criteria they were looking for in their custom growers and grass finishers.

1. Multiple years experience with MiG.
2. Willingness to grow and utilize cool-season and warm-season annuals to extend the harvest window and increase seasonal gains.
3. Ability to create a planned forage sequence of various perennials and annuals capable of creating high gains over an extended period of time.
4. Ability to provide strategic irrigation.
5. Planned contingency program for weather emergencies.

"We now want our graziers to have a high quality stored forage available 365 days a year," Williams said.

"Also, you need to base your stocking rate on your worst month of the year and not your best."

He said he knew they were asking a lot of graziers but that they were willing to pay them well for their trouble.

He said Tallgrass had devised a sliding scale that paid more as the animals gained weight. They also pay more for heifers than for steers.

A major problem is the poor cattle handling facilities on many graziers' farms and ranches. "You absolutely have to be set up to weigh the cattle periodically. No one can eyeball how well an animal is gaining."

Not all animals must be weighed but a 20% representative sample must be weighed every 60 to 90 days. A minimum average daily gain of 1.5 pounds per day from arrival to harvest is the contract target gain.

Crop residues can be used with dry cows but not with calves and yearlings.

No grain, corn silage, animal by-products, fish by-products, bakery by-products, candy or potato products are ever allowed under Tallgrass's protocol, and all supplements must first be submitted for approval before using. A signed affidavit testifying that the protocol has been followed is required of each grazier.

Williams said finished animals absolutely had to have a minimum of 3/10ths an inch of fat cover to prevent cold-shortening in the chiller. The minimum acceptable degree of marbling is 3.5% backfat or USDA High Select.

Interestingly, thanks to a stringent pre-selection program, Tallgrass actually has more cattle grading Choice than the industrial feedlot average.

"We run about 60% USDA Choice, versus 49% for the grainfed average. Also, we run about 2.5% Prime grade, which is exactly the same as with grainfed," Williams said.

Current carcass yields range from 59 to 63% with average carcass weights of 625 to 725 lbs.

"High carcass yields come from animals with deep bodies and fullness from front to rear. We have no use for tall, lanky cattle," he said.

Life

When I first discussed doing a book on ranch financial issues, my wife, Carolyn, insisted that it include lifestyle issues as well as purely financial ones. She argued that the business side of ranching was so intertwined with family, emotional and cultural issues, which often over-rode purely financial ones, that to ignore them was folly.

And, as usual, she was right.

Whereas, in most businesses lifestyle is a *result*, in ranching the desired lifestyle most often trumps all business decisions. From my 30 years of experience as a change agent, I have learned that very few people respond to purely economic arguments. Even things that are wildly more profitable than the status quo are rejected if they run afoul of a rancher's cultural predisposition.

If you want to pursue something more than an "average" operation, you are going to have to withdraw from the dominant culture and seek out like-minded people like yourself wherever you can find them.

Just realize that nowhere are there great gobs of such people, so don't go looking for big crowds.

If you find small numbers discouraging, consider that one percent of the American people earn about 37% of the total income, 10% earn 67% and 25% earn 84%. Do you want to be with the huge crowd of 75% of the people who earn 16% of the income or the top one percent?

Who you hang out with largely determines your destiny. These statistics should prove that if a lot of people are doing something they probably aren't making much money at it.

Wealth comes to the few who have the guts to defy the majority opinion and the spending discipline to buy assets rather than depreciating toys.

Wealth is found on the sunny side of the street

Do you remember telling ghost stories around the campfire when you were a kid? Remember how scared you got even when you were telling a ghost story you just made up? You knew it wasn't true but it scared you anyway. This phenomenon is called cognitive dissonance.

In other words, your subconscious believes what it hears your mouth saying even when you consciously know you are just telling a tale. It is easy not to believe other people's ghost stories, but your subconscious believes you. You feel scared and act scared because deep within you are scared.

The next time someone starts to tell a ghost story about some dire thing that is about to befall us in the livestock business ask him, "Yeah, but what's the opportunity in that? Who's going to make the money in that scenario? How can we make that money?"

You'll be surprised at how many people this will infuriate. In fact, usually only one person in a hundred will want to play this opportunity exploration game with you. Why is that?

The other side of cognitive dissonance is that if I believe in the reality of an opportunity, I feel obligated to pursue it. Therefore, the way I protect myself from this feeling of obligation is to refuse to believe in the opportunity.

How many times have you heard, "Yeah, well you might can do that where you come from, but you can't do it here."

That's the sound of a person protecting himself from the tension opportunity obligation brings. This person is not telling the truth because he doesn't know if he can do it or not. He is just telling a ghost story.

However, because he genuinely feels scared when he tells it, it reinforces his belief that it must be true.

You cannot argue him into seeing your point of view.

> *The reason it is so hard to follow the leader is that the leader got there by doing something remarkable. And that remarkable thing is now taken and is no longer remarkable when you do it.*
>
> *Seth Godin*
> *The Purple Cow*

You cannot prove it with research or data. In a twisted logic, he feels that his own lack of success validates his point of view. In other words, he has a vested interest in failure. How many times have you heard people say they have tried something "to prove it wouldn't work." Where's the glory in that?

I read that people who sell commodity-priced goods tend to develop conspiracy theories and bogeymen to rationalize their losses. Because there is always a winning side to every trade, it is comforting to think that someone has stacked the deck.

It is unfortunate that negative people can usually dominate a group. Negative thought is such a strong de-motivator that most of us cannot stand up to it. One negative person can totally sap the energy of a roomful of people.

This is why people who decide to do something different in life have to withdraw from the crowd until they have proven to themselves that they can do it.

If your goal in life is to create wealth, you absolutely have to keep a positive outlook in seemingly negative situations. Wealth comes from exploiting opportunities. It is only within problems that opportunities lie.

Therefore, problems are the opportunities. There are no self-made wealthy pessimists.

The early 1980s was a time of widespread bankruptcy

and despair in the North American agricultural community.

I had gotten wrapped up in the overwhelming sense of doom and gloom and had written that the situation was so bad that "everyone was going bankrupt."

A young farmer from Florida called me soon afterward and forcefully declared that not "everyone" was going bankrupt in agriculture. In fact, he said that as a full-time farmer he was doing extremely well financially.

"Why don't you concentrate on writing about the few farmers who are doing well rather than the ones who are failing?" he asked.

This led to a rather long relationship with this particular farmer. I went down to visit him to learn his secrets and it was from him that I first heard the hypothesis of the absolute necessity for an "unfair advantage" to succeed in a commodity-priced business.

He explained that farmers had to learn to "farm the weather" and exploit the inherent advantages in their location. In Florida, he said their unfair advantage was a very mild winter and he had designed his farm to exploit this "unfair advantage."

"You must learn to compete from your strengths," he said.

Bud Williams said the biggest thing wrong with telling

The weaving together of business and family plans represent a special challenge for the family business, because it means that the business and family plans are highly interdependent.

The business plan requires the family to determine the extent of its commitment to the farm. That commitment depends upon the prospects for the farm that the family planning process reveals.

As a result, the family cannot separate strategic business planning from family strategic planning. It must undertake both in a connected and simultaneous way.

John Ward
Family Business Review

people to identify and exploit an "unfair advantage" is that it is inherently "unfair."

He said the principle of "fairness" is so ingrained in most people that if a proposed solution is not universally applicable it has become politically incorrect to even discuss it. It will frequently be shouted down with questions similar to the following:

> *Capital lets you have an aura of safety around yourself and allows you to enjoy yourself and not live in a cage of fear. And the first step and the last step is self-discipline.*
> *Ben Stein*

"If we go seasonal, who will make milk in the summer/ winter?"

"If we start grass feeding, who will keep the corn farmers going?"

"If we switch to stockers, who will produce the calves?"

"If we calve in the summer, who will calve in the fall and winter?"

The only possible answer to these questions is "Other people."

Bud Williams said that to become wealthy you must become comfortable with the moral "rightness" of getting more than your fair share.

Regardless of the commodity produced or the commodity's price, he said there are always 15 to 20 percent of the producers doing exceptionally well.

If you want to be wealthy it is those people you should be watching and studying, and only those people.

"Most ranchers prefer to copy others because they lack confidence," he said.

"And that's okay as long as you only copy people who are really making money."

He said the reason these people are doing well is because they have developed some bit of knowledge and are exploiting it.

The beauty of knowledge is that once it has been obtained it has no on-going direct cost. It is also invisible to your competitors and they have a tough time copying what they can't visually see.

All change starts with innovative few

Jim Gerrish, author of *Management-intensive Grazing*, once pondered why it was that a rancher feeding hay for five months of the year couldn't see that his next door neighbor was feeding none?

I think I may have the answer.

I found it in a 1962 book called *Diffusion of Innovation* by Everett M. Rogers.

This book is a study of the extension efforts to get farmers to switch from open pollinated to hybrid corn in the post World War II era.

This study forms the basis of all modern marketing thought and is still widely studied as a guide on how one creates technological change.

While this study was on farmers, subsequent researchers have found that it applies to all new technologies.

A more recent study of the diffusion of computers found the process followed exactly the same diffusion path as that outlined by Rogers.

However, I will confine my observations to farmers as Rogers did.

Rogers found that all technological change starts with a small group he called the "innovators."

This group is no more than 2.5% of the total group of farmers and ranchers.

This is a venturesome, restless group who like new things because they are new. And, this is the **only** group of farmers who truly like to learn new things.

Marketers have discovered that trying to take a radically new idea to the larger non-innovative group is a total waste of time and marketing dollars.

Everything new has to start with this small group.

Rogers found that innovators tend to be wealthier and far better read than most farmers.

They like to travel widely and, above all else, like to visit with other farmers like themselves.

As a result, they form many long-distance relationships with other innovators.

> *Things that alter the world invariably enter by the side door. A small group of pioneering consumers, paired with a pioneering vendor, act as flag bearers for the rest of the world.*
>
> *Tom Peters*

They can understand and apply complex technical knowledge and can adapt a new idea from just reading about it in the trade press.

(Only eight percent of the human population can build something by reading a set of plans.)

They are willing to accept an occasional bloody nose from pioneering a new field and do not quit when things do not work out exactly as planned.

They tend to specialize in one type of enterprise. (All pastoral or all arable.)

They are less dogmatic, less rigid, more rational, and their neighbors think they are totally insane.

As Joel Salatin's neighbors have defined Joel, the innovative farmer is seen by his farm neighbors as a "lunatic farmer."

And, a lunatic is not seen as a role model.

As a result, what the innovator does on his farm is literally invisible to his neighbors.

This is true even if the innovation is producing visible wealth in the form of new homes, a growing herd size and expanded acreage.

The best an innovator can hope for is an admission from his neighbors that what he is doing works for "some people."

However, the normal reaction to unconventional success is the old it-might-work-there-but-not-here syndrome. "There" can be, as Jim Gerrish said, as close as the next farm or ranch.

Psychologists have found that humans deal with today's information overload by filtering out all information not deemed relevant to them.

For example, few Republicans hang around to hear the Democratic response to the State of the Union address and the reverse is also true.

Most of us don't want "fair and balanced" news. We want it to fit our existing mindset.

The traditionalist farmer is not moved by economic arguments generated by people from outside his peer group.

The sad truth is that the vast majority of farmers prefer to fail conventionally rather than to succeed unconventionally.

Most people feel safest in staying with the large group even if they don't particularly like where the group is going.

The Christian philosopher, Anthony De Mello, said your first goal in life should be to teach yourself to swim so that you don't drown with your friends.

Of course, the "swim" here is a metaphor for all of life's survival skills.

Rogers found that some communities had greater tolerance for innovation than others. The Deep South and parts of the Midwest have next to none.

It is very, very difficult to be more innovative than the community in which you live.This is why poor parts of the country tend to stay poor regardless of the health of the national economy. These areas purposefully drive away the very people who could create a better economic environment.

However, the antipathy traditional farmers feel for innovators is mutual.

Rogers found that 89% of innovative farmers see

themselves as different from their neighbors.

While they might be geographically a part of a community, the innovator was never philosophically in tune with it and generally, for the most part, lives apart from it.

Innovative farmers think their traditionalist neighbors are a bunch of sticks in the mud and actually seek to have as little interaction with them as possible.

They are only influenced by farmers as, or more, innovative than themselves.

They distrust university and private extension efforts because they are not peer generated.

Innovative farmers actually prefer to get their information from outside their own community; whereas, the vast majority of farmers distrust anything from "away."

> *I had a student who told me, "If I ever make $5 million I'll call it quits."*
>
> *I told her, "That's why you will never make $5 million."*
>
> *People who are successful in life do not wake up one day and call it quits. The ones who want to call it quits are the ones who will never have anything to quit in the first place.*
>
> *Keith Cunningham*

Interestingly, Rogers found that no farmers, innovative or traditional, referenced what a farmer smaller in acreage than themselves was doing as applicable or worthy of study. Everyone preferred to learn from someone larger and more successful than themselves.

No doubt this is the true source of the "get bigger or get out" train of thought. Like football players we all seek to play at the next level.

Not surprisingly, people who come to agriculture from other careers tend to be drawn to doing innovative new things because they have no financial or emotional stake in the status quo.

Ironically, innovative farmers are not as independent in

their thoughts as they think they are.

Innovators tend to rather quickly coalesce around generally agreed upon principles of the innovation they are pursuing.

As Rogers said, "They are in step, but to a different drummer."

This gap between the innovator and the traditionalist has been defined as being more akin to a "chasm" than a small ditch.

For an innovation to become widely adopted, it must somehow find its way across this cultural divide.

Rogers found that within the traditionalist majority there was a small group of farmers who had learned to watch what the innovative farmers in their community were doing.

They, unlike the rest of the traditionalist farmer community, are always watching over the fence.

Once the innovator had perfected an economically viable "whole," they quickly adopted it.

And because these "early adopters" were a part of the larger traditionalist local community, they were seen as suitable role models.

This early adopter group formed the critical bridge across the "chasm" that allowed the innovation to spread outside of the tiny innovator community.

The innovators and the early adopter group form approximately 15% of the total farming community. Interestingly, this percentage is almost exactly the same as the number of farmers who earn an upper class income (16%) from agriculture.

Tools and toys

I got a phone call from a very upset reader one morning. "My neighbor spends all of his spare money on bass boats and fishing rigs. Is there anything wrong with that?" The belligerence in his voice was crackling the phone wires.

I told him, no I didn't think there was anything particularly illegal or wrong about that.

"Well, then why the (blankety-blank) can't I spend my spare money on what makes me happy?"

And what would that be?

"A really nice new roundbaler and a mower-conditioner."

Suddenly I knew why he was upset. I had written a recent column about avoiding "rust, rot and depreciation"and the foolishness of putting one's hard earned money into little-used machinery. Unfortunately, his wife had read the article and took him to task about his growing machinery collection. I tried to explain that buying a bass boat was not a business decision and that farm machinery was, but I knew it was hopeless.

Separating business decisions from lifestyle decisions is very hard in agriculture. This is because a lot of the legitimate tools that we buy with before tax money like horses, shotguns and ATV's also have recreational uses. It is hard not to slip over the line from buying legitimate, needed, tools to buying mechanical farm toys to play with.

We are like European shopkeepers who live above their stores. It is hard to draw a hard line between where work stops and family life and recreation begins. We can go fishing in the ponds we build for stockwater. We can hunt rabbit and quail in our pastures. And, because we like horses, we aren't too concerned with figuring out if they are really needed. We really like our work, so we aren't too concerned about how much we pay ourselves to do it.

All too often the family and farm expenses are drawn from the same pot of money. Soon no one has a clue as to whether or not the farm or ranch is making money except that the wife has to go to town to get a job. After a few years, she begins to suspicion that she is working to support both the family and the farm but she has no way of proving it because no books are being kept.

Most men don't go to the doctor because they don't want to know if there's anything wrong. Financial records tend to be looked at in the same way for the same reason. Unfortunately, its virtually impossible to improve what you can't measure.

In my 30 years as editor of the *Stockman Grass Farmer*, I know that urging farmers and ranchers to keep a legitimate set of books is probably one of the least popular speeches you can give. We are much more interested in hearing about ways to increase the top line than figuring out whether or not there is a bottom line.

However, since this is a book on finance I will tell you that hiring a good bookkeeper will probably be the best money you ever spent. You need to separate the farm and ranch business from your personal life and separate what are toys from what are tools.

Draw a monthly salary and pay the family bills out of that. Put a pencil to how much money your wife actually brings home after taxes and outside child care expenses. I guarantee you in most cases, she could make more net income with the egg money from a few dozen free-ranging hens.

Let your children earn their own money with alternative species of livestock or set up a roadside stand and let them sell the surplus from the garden.

If your children work in your primary enterprise, pay them! The labor of children has a real value on a farm. You'll be surprised at how much more interested they are in what you are doing.

Figure out how many more cows you could buy with the money you've got tied up in little-used machinery and then sell the machinery and buy those cows.

And, then if some day you really want a new roundbaler and mower-conditioner, buy it with the money from your personal mad money account just as you would a new bass boat or any other toy.

The bottom line is that a profitable ranch requires a huge amount of spending discipline. Keep your toys and tools separate.

Avoid the treadmill, let the animals do the work

In the world of small business — and particularly with the self-employed — scalability can be defined as the ability to produce more output without an equal amount of effort being expended. I define this as "Bucks per bucket of sweat."

For example, in a publication such as *The Stockman Grass Farmer*, an increase in circulation requires far less increase in staff labor than an increase in advertising. One more subscriber means that the press runs a few seconds longer and requires no additional staff labor costs. However, an increase in advertising requires additional editorial pages which increases both layout and editorial labor and costs.

Enterprises that have near a one-to-one ratio of labor input to output are often referred to as "treadmills." You can run harder and make more money but you cannot lessen your effort without an equal diminishment in income. Any work that is charged out by the hour — whether it is a psychiatrist or a day laborer — soon runs into the fact that the amount of time in the day becomes the primary limiter to the amount of money you can earn.

This is also true for self-employed people like farmers and ranchers. Enterprises that cannot be time leveraged quickly top out in the income they can generate because you run out of time and/or energy.

While many of his customers suggested that he just hire cheap labor and keep expanding all of his activities, he said his dream was to create a profitable "family" farming operation.

"We tried empire building and it is just not for us."

Joel Salatin defined farm scalability in terms of size neutral and non-size neutral livestock operations. For example, he said pastured beef, and pork were size neutral, meaning that they can increase in volume with only minute increases in labor. As an example, he said moving 500 steers to a fresh paddock is no more work than moving five. Thanks to the use of low-labor self-feeders, pastured hogs are a check-every-couple-of-days kind of enterprise.

However, he said broilers, laying hens and rabbits are not size neutral. Each additional animal requires a proportionate increase in daily labor.

"The 56th dozen of eggs cannot be put away any faster than the 55th," he said.

"We were carrying feed and water, but not sitting on the edge of a poultry shelter to revel in the sun's radiance dancing off dew-bedecked clover leaves. Our success was unraveling our lifestyle."

Consequently, he has expanded egg production by recruiting his female neighbors as contract egg producers.

"Women like laying hens. We tried using men for several years and they just weren't responsible enough."

In 2006, Joel was looking for a young person to take over operating his pastured broiler operation on an income split basis.

"We like the symbiosis of getting the chicken's manure on our land but the chickens are very time consuming," Daniel Salatin said.

The Salatin's are currently putting a lot of effort into expanding their grassfed beef operation. Joel has moved his cow-calf herd and stocker operation to leased land and now only does the final grass "finish" on the home farm. He has been able to greatly expand his grazing acreage through his reputation for creating "pretty properties." Consequently, he has been able to add several hundred acres for little more than the land taxes.

He is also expanding the farm's acreage in horticulture.

He currently has blackberries, apples and mulberries and is adding pears and seedless grapes. Fruit does well on steeply sloping land unsuited to his other enterprises.

Joel said grazing poultry beneath fruit trees offers a unique symbiosis. The poultry help protect the fruit trees from disease and the fruit trees help protect the poultry from aerial predators. He said turkeys make the best orchard gleaners because they do not dig bare spots in the grass the way chickens do.

Also because of their large size, turkeys return more to processing labor than chickens and can actually get a considerable portion of their diet from grazing. As a result, turkeys are much more scalable than chickens.

If you feel the same way as Joel about hiring outside labor, your primary enterprise absolutely has to be one that is largely size-neutral if you want to avoid the treadmill as you increase volume. As Joel has found, self-feeding animals are an excellent choice. A well-designed ruminant operation can be a very low-labor input operation regardless of its size.

Working smarter rather than harder is the way to wealth. Don't put your nose to the grindstone. Follow Joel's example and stick with enterprises that give you some time to dream.

Employees are the only way to have more leisure time

A major problem we have in North American agriculture right now is that there is no way in for young people. Either your own or someone else's.

This is a real problem because few people are mature enough to self-motivate and manage their own affairs until they reach the age of 30. Most of us need to start out working for someone else.

A lot of ranchers are rocked on their heels when they hear Bud Williams' views on ranch labor productivity. Bud said the biggest problem with ranch profitability is that most ranches have too few people on them. He thinks that good people, trained properly, are a major ranch asset.

He said for 50 years we have replaced people with depreciating machinery in hopes of being able to buy ourselves more leisure time. Unfortunately, machinery is only cost-effective when it allows you to do more work - not less.

The only way you are going to have more leisure time is through OPL or "other people's labor."

He said the big fallacy in the idea of one person trying to run a ranch is that no one is, or can be, good at all things.

"We all have our strengths and weaknesses," he said. "You can't build a ranch around your weaknesses. You've got to hire other people who offset your weaknesses."

I recently read an explanation of why money is important that echoes Bud's thinking. It said that money buys you the time to do what you love by allowing you to pay other people to do what you hate.

I am sure in your operation there are aspects of your job you love and aspects that you hate. I believe the reason so many of us get "cowed out" in our 50s is that we are forcing ourselves to do those things we hate doing rather than spending all of our time on those things that we love.

> *You have to teach your employees how to be profitable. Chores are created to consume excess time. New employees copy what the old employees did. No one is thinking about what really needs to get done.*
>
> *Bud Williams*

A lot of people are shocked when they hear that our small publication now has 14 employees to do what I used to do all by myself. And yet, my paycheck is many times what it is was then and today I get to do only those things I like to do!

Bud Williams is right. Good employees don't cost. They pay.

Being able to do only what you like to do is the big difference between being self-employed and owning a business. Making your ranch into a business starts with your first employee.

The trap of self-employment is that your income is directly related to your own personal labor. If you take a vacation, so does your pay. However, hiring an employee is a big leap for most self-employed people.

Like most new things, this is best done on as small a scale as possible until you become comfortable with it. This means starting with part-time employees.

Bud Williams said the problem with hiring a full-time employee is that you feel the need to fill up their day with work. On a ranch this is very difficult to do, so we invent chores.

Chores are invented because no employer can stand to see an employee sitting still and doing nothing.

However, unproductive make-work can actually take up

so much of the employee's time that the truly productive work doesn't get done.

So, we hire another employee to get the needed work done but add some chores to fill out his day. Soon we need another employee.

And another. And another.

Now, we need someone to watch all of these people to make sure they aren't goofing off in the haybarn and these kinds of people don't come cheap.

What is overlooked is that 90 percent of what these people are doing every day doesn't really need to be done. The best way to avoid this wage expansion trap is to hire people to do a job rather an eight hour day. This requires you to first identify what it is that you want done.

On a grass-based farm, a lot of what we need done is baby-sitting type of work.

Are the waterers working properly?

Is the fence hot?

Do the cattle need to be shifted to a new paddock?

The vast majority of work on a grass-based farm or ranch is not overtly physical. Most of it requires a responsible employee rather than a muscular employee. Once you realize that, your potential work pool expands tremendously.

> *Services are worth plenty when you have more money than time.*
>
> *Thomas A. Stewart*
> *The Wealth of Knowledge*

Break what you do every day into specific jobs.

The first thing you personally want to get rid of are the baby-sitting type of jobs that anyone could do.

What you are trying to do is to hire yourself more thinking time. As Stan Parsons said, you need to spend some time working "on" the business rather than spending all your time working "in" the business.

Figure out what getting rid of a particular job would be worth to you and then hire someone to do this one specific job.

Tell them when and how you want it done, but also tell them once they are trained you expect them to do it without your having to be there.

If they can figure a way to do this job in five minutes a day, that's fine with you. In fact, that is even better as they won't be getting in your way or taking up your time by talking to you and trying to be "neighborly."

While this kind of job may sound less attractive than a full-time job, there are actually far more people looking for this kind of flexibility and freedom today than there are people looking for a full-time, minimal wage job.

These people are not young people. They are retired and semi-retired people in their early to late 50s and early 60s with good educations and work habits.

Keep in mind that you are not looking for someone entrepreneurial like yourself. If you hire someone like yourself, you will just be training a competitor. You are looking for someone who likes to be an employee.

> *Since we do not revere the elderly, we create the illusion we can stay young forever.*
>
> *Clotaire Rapaille*
> *The Culture Code*

Many government employees are allowed to retire after 30 years of service regardless of age. This means they can often retire in their early 50s. This group includes military, police, fire fighters and schoolteachers.

Since these people have a pension, they neither need, nor want, full-time employment. However, they do want to get out of the house, have a daily routine and some responsibilities.

Many of these people have spent their whole lives working in a rule-bound world where inter-department politics determined who got ahead. The idea of just being given a job to do at their own pace and recognizance can be totally liberating and exciting.

The other large pool of people looking for this kind of

work is women with school age children.

If your work can be structured so that it can be done between nine and two and these women can have the flexibility of being able to miss a day if a child is sick, you can find excellent, well-educated, highly responsible people.

Personally, I think these kinds of women employees are tops.

> *Success is about who you are, not what you have.*
> *Anthony DeMello*

Most women seek perfection in all that they do. They want to become really good at whatever task they take on and do not need, nor want, direct supervision.

However, because they like feeling competent they do not like sudden changes in their job duties.

I have learned that any change in job routine needs to be accompanied by a thorough explanation of the economic necessity for this change. Female employees want to know this is not just a snap decision that has been poorly thought through.

(They seem to think males do this a lot.)

Also, you must realize that their families take precedence over their jobs. If Junior has the flu, you had better be prepared to shift the steers that day yourself.

However, she will always call and tell you her problem rather than just not show up as male casual labor frequently does.

Women tend to be more sensitive about animals' pain and stress than men and seldom fall in love with tractors or large trucks. As a result, they are particularly well-suited to a grass farm or ranch.

We have found a good way of finding stable, responsible women employees is through ads on church bulletin boards, or by talking to friends to see who they know.

The whole trick in learning to like having employees is to always hire responsible people with good work habits.

Good ranch employees do not have to be young, male,

nor full-time, to be highly productive.

Now, I'm sure there are people who take great offense at the idea that children should work for their keep. However, Joel Salatin makes the case that on a pastoral farm there is very little difference between child's play and work in many activities.

For example, Joel said that he recruited his son, when young, into helping him dig pasture thistles by putting it in the context of a war on the nefarious "Thistle Nation."

Similarly, when Joel would have to go out of town for several days he would build several days worth of temporary paddocks so all Daniel had to do was to open a gate each day. This is what I mean by casual labor. It's not physically hard work but it is necessary work.

It is Joel's argument that a well-designed grass farm is and should be child-friendly for both work and play. Joel said the critical element in child's work is that it should be of short duration. The other part of Joel's hypothesis is that children should be paid, and paid well, for their labor.

> *What's good about being a farm kid:*
> *1. Farm kids have a real sense of their own significance. They feel they matter, and they do.*
> *2. Farm kids tend to be more confident and responsible.*
> *3. Farm kids and their parents share more activities.*
> *4. Farm kids can adapt well to changes in economic circumstances. They know how to cope.*
>
> *Glen Elder, Jr., Sociologist*
> *University of North Carolina*

His guiding principle on child's pay is "give them what it would have taken you in today's dollars to feel so good about the work that you'd want to do it again."

He would rather deny himself an income, if things are tight, than make his children think the farm can't pay them anything. Once a child reaches adolescence, Joel believes that if he is doing a lot of the work on the farm, he needs to be paid at least $10,000 a year.

"Remember, they are rubbing shoulders all the time with people who plan to earn a lot more than that in city jobs. The way to head them off at the pass is to show them the potential early on, so the stars in their eyes are focused on the family business, not somebody else's business," he said.

> *Age wrinkles the skin. Quitting wrinkles the soul.*
>
> *General Douglas MacArthur*

He said it is very important that your children see the farm as a viable vocational option early in their life if you expect them to stay.

American business guru, Peter Drucker, said the only business structure that has shown it can last over time is a family business.

Here in the USA only three to four percent of private family businesses make it to the fourth generation. However, in Asia and Europe there are family businesses that have been ongoing concerns for hundreds of years.

A good book about long-lived family businesses is *Centuries of Success* by William T. O'Hara. This book studies family-owned businesses all over the world that have survived at least 200 years.

Needless to say there are very few New World companies that can meet this criteria because of the young ages of our countries. As a result, this book has a definite "old world" feel and attitude that may be unsettling to egalitarians.

Peter Drucker said that for a family business to survive it must be structured to be a business first and a family second. This book agrees with Drucker's observation.

While each business profiled in the book has unique aspects to it, a common theme among them is that inheritances must be designed to avoid fragmenting the production base of the business.

The author said this need to maintain the whole is the single factor that all family businesses have in common.

In Japan and Europe where primogeniture was — and is practiced — there are many multiple-century-old businesses. Primogeniture is the transfer of the business in whole to the oldest son.

> *Retirement is simply doing what you want, when you want to do it. Why not fall in love with what you are doing and retire into your job?*
>
> *Keith Cunningham*
> *keystothevault.com*

In Japan, if a man does not have a son he adopts one of his sons-in-law as his business heir. This son-in-law must renounce all inheritances from his birth family and change his name to that of his wife's family.

In other words, he hitches his wagon to his father-in-law's horse and gives up any financial tie or backup to his birth family. This insures that his financial success is tied solely to his father-in-law's business.

Under primogeniture, part of the inheriting son's responsibility is to make sure that his brothers and sisters are financially provided for.

He is to encourage his brothers to start new businesses within the family "whole" which they can manage without interference and from which they can pocket the majority of the income.

And, he is to make sure his sisters find "suitable" husbands of the right financial status and are otherwise always provided for.

Needless to say a lot of responsibility goes with being the oldest son. Some European heirs refuse it.

In Asia, the financial benefit of being a part of the family "whole" is designed so that losing your familial connection would be financially devastating. As a result, the family elders can insist that the children never do anything that would bring dishonor to the family name.

For example, any member of the family can be expected to hold and protect money for any other member without fear of it being stolen or misused.

Again, this honesty is enforced by family expulsion if violated.

In Asia, such "family" connections exist to eighth cousins. By spreading cousins around the world, these family businesses were able to build international trading businesses and to protect the family "nest egg" from war, currency debasement and political unrest.

Now, these ideas are totally foreign to most North Americans. Only one-third of our family owned businesses transfer one generation. Most die with their founders.

The average life of a "successful" American business is 24 years or roughly the "useful" working life of its founder.

(By useful I mean from age 30 on. Very few people are emotionally mature enough to start a business younger than 30.)

The American attitude is "I got mine, son, and I hope you'll go and get yours too."

This stems from the predominance of Anglo-Saxon culture in the USA which emphasizes an "every man for himself" approach

Family business systems:
1. The Family Circle of Love — The family takes what it gets and loves you for it.
2. The Business Circle of Competence — You must prove you can add value to join.
3. The Ownership Circle of Control — He who owns the gold makes the rules.

The senior generations have to set up the ground rules about how you move into the other two circles and how you will be paid, what will get you fired, and when and how you can attain ownership of the gold.

These discussions should be held when your children are in High School.

Linda H. Smith

to life. In contrast, Germanic culture emphasizes the need for the father to provide the means of living for his sons.

In North America, this cultural difference can be seen strongest in the Amish where the father is expected to provide his sons with a paid-for farm.

> *How do you know when you are rich? No debts, no mortgage, no overdraft and you are able to pay all of your bills by return mail.*
>
> *Paul Johnson*
> *Forbes Magazine*

No doubt, our short-term oriented, Anglo-Saxon culture serves us well in high technology. If your goal was to build a business for your son, you surely wouldn't go into high-tech where business lifespans are normally less than five years.

So, what would you go into?

Surprisingly, O'Hara said agriculture is not a bad choice.

The longest lived American business is 365-year-old Shirley Plantation in Charles City, Virginia. This large acreage farm leases its crop acreage and now concentrates its management on agritourism and producing grassfed "helal" sheep and turkeys for Islamic butchers in the Washington, D.C. area.

Thumbing its nose at Europe's male primogeniture, Shirley Plantation is currently headed by a woman.

Author O'Hara found an over-abundance of long-lived family businesses in the wine industry.

This point intrigued me as many people have noted that the emerging grassfed industry most closely resembles the wine industry.

The common factor in the wineries studied that have lasted for centuries was a total concentration on producing a product that could command a wide financial margin per unit of production.

This wide margin allowed the family business to stay small and self-managed.

Family businesses that go into anything where scale is

an important competitive factor almost always lose out to public corporations and their access to cheap capital.

O'Hara found one thing that could provide this wide financial margin year-in and year-out was a product whose quality was so exceptional that consumers willingly paid an above-average price for it.

No shortage of opportunity for your children

Joel Salatin said on a well-designed farm nothing is wasted. For example, the sawmill's slabs and trimmings from the forestry operations provide the wood for heat and hot water for his and his mother's house via an outdoor control-burn, wood-fired boiler. The boiler's ashes provide calcium and potassium-rich fertilizer for the pasture and a mite-fighting dust bath for the chickens.

Joel estimated the "income" to the farm from feeding eight people and heating four houses to be worth at least $30,000 a year in what would be after-tax income for city people.

He said the farm served as his family's home, livelihood and primary source of entertainment. He estimated that his family spends no more than $15,000 a year on non-farm related expenses. This has allowed for a large financial surplus that has been reinvested into the farm each year in tax-deductible infrastructure development and improvements.

The drought of 2002 proved the value of pasture irrigation but also proved that the farm also needs far more upland irrigation water catchments, and were budgeted to be a major development expense for several years.

In his late 40s, Joel turned the operation of the farm over to his son, Daniel. He said any future "holons" would come from his children's or grandchildren's particular interests and not his.

The best way for a child to create a sense of independence and success is to be able to learn something that Dad doesn't know. And, because it has no direct cost to Dad's primary enterprise Joel argues the child should be allowed to have all, or most, of the income it generates.

On Joel's farm, broilers are raised in floorless shelters, which are moved across the pasture after the cattle. The broilers fertilize the pastures for the cattle with their manure. The cattle shorten the grass for the chickens.

The cows drop their manure which incubates flies and they uncover grasshoppers and crickets with their grazing. This provides a feed resource for free-range hens.

Following the cows through their rotation is a two-house train of portable laying hen houses that Joel calls eggmobiles. Nearly 1000 hens free range out from the eggmobiles, to eat fly larvae and incorporate the fertilizer into the soil surface.

Along with the pasture sanitation, the chickens harvest the newly uncovered insects turning what was a problem for the cattle into an economic resource. Joel said the free-range hens have completely eliminated heel fly warbles in the cattle.

The cows go across the paddock six times a year depending upon rain and hay schedules. The eggmobiles follow the cows in each pasture rotation. The broilers go across once and the

> *Most of today's farmers entered in the boom times of the 1960s and 70s but their children have seen nothing but hard times. These children were typically born in the late 1970s and "cut their teeth" on the farm crisis of the early 1980s. If financial strain has made parents bitter and angry about farming, their sons aren't apt to stick around.*
>
> *Kids who stay on the farm have parents who perceive farming as a good way of life and a good way to make a living. Kids who leave have parents who would themselves leave farming if they could.*
>
> *Glen Elder, Jr. Sociologist*
> *University of North Carolina*

turkeys go across once. All grain used in the poultry operation is imported from off the farm and is the only fertilizer input the farm is currently using.

Now let's look at the income involved here. Keep in mind all production is sold directly to consumers.

The broilers will gross about $3000 an acre and net $1500.

The cows will gross about $300 an acre and net $200.

The eggmobiles will gross $500 an acre and net $200.

The turkeys will gross $1500 an acre and net $800.

Add it all up and you have around $2800 an acre in gross margin in eight months of the year.

Joel guesstimates that his mostly forested 500 acre home farm could support at least 20 salaries at maximum utilization!

Here are the enterprises and holons that Joel has in operation or has considered for his farm.

- Grass finished beef
- Pastured pork
- Pastured laying hens
- Turkeys
- Firewood and timber forestry
- Sawmill
- Wood working shop
- Rabbits
- Sheep
- Orchard
- Grapes
- Marketing/promotion
- Delivery van
- Vegetables
- Small fruits
- Farm and ranch vacation cottages
- Aquaculture (caged fish production in farm ponds)
- Hoophouse vegetables
- Mechanic/fabricator

■ Accountant/bookkeeper
■ Trail rides
■ Chef for info-tainment dinners
■ Inspected kitchen — pot pies, bakery etc.
■ Greenhouse

Joel said that most of these enterprises do not require an increase in land area to be put into place. Any one of them offers "a way in" for a young person.

Non-production enterprises

Value adding activities can create even non-farm job opportunities for children as well. Some of them Joel mentions are:

■ Make ice cream on a dairy
■ Make ground flour from grain
■ Turn the grain into muffins
■ Make apple juice from apples
■ Bed and breakfast to add value to eggs, pork and grain products.
■ Tan hides
■ Make gloves and moccasins from tanned hides

Joel said it is important for parents to keep presenting untapped opportunities to their children.

"If the children sense that the cornucopia is going to run out, they will begin looking elsewhere for new fruit," he said.

Joel believes that rather than saving for retirement parents should put their money into funding their children's new enterprises.

"The money pumped back into the profitable business via incentives to the young will generate far more equity than you can by keeping the children as slaves and dependents as long as possible so your accumulation at the top can be higher," he said.

"$10,000 given to a 20 year old can fund a business that can last a lifetime. $10,000 given to a 50 year old will probably fund a two week cruise in the Caribbean."

> *First, recognize that the farm business and the farm land are two different considerations. The only children who have a right to the farm business are those who work in it.*
>
> *Leaving the farm's land quality divided among your children means the parents want to see the business continued. If not the land should be sold while the parents are alive and a cash inheritance provided.*
>
> *It is the parent's responsibility to build a business structure that will survive them.*
>
> *Australian Farm Journal*

In return for getting their parents' wealth before they die, the children should be expected to take care of their parents when they are old.

Joel also believes that the child who stays on the farm should inherit it. It should not be equally divided with the siblings who left for jobs in the city and contributed nothing to it.

While many of Joel's ideas sound radical, most are really a return to a traditional family business paradigm. In Asia, family business structures encompass far more than the primary family and extend to fairly distant cousins as well. This has allowed the Chinese, for example, to develop family businesses that are global in scale but that pretty well follow Joel's outline.

Mainstream business guru, Peter Drucker, said the only two business structures that have withstood the test of the centuries are to:

Run your family like a business.

Run your business like a family.

One point Joel makes that I absolutely agree with is that if you want to have your child stay on the farm you had better watch your attitude.

In his book, *Family Friendly Farming,* Joel said you owe it to your children to add irrigation to your farm or ranch. I agree. I know I would have gladly signed away any rights to an inheritance to have gotten my rancher father over his constant weather-related blues.

Remember, your children want you to be happy every bit as badly as you want them to be happy.

One point, my father agreed with Joel was the necessity of "hardening off" your children before they leave home and have to face the rigors of the world. However, Joel's recommended methods and my father's were considerably different.

When I was 15 years old my father called me into his ranch office one day in late May. He had been concerned that at the rate I was getting beat up by bullies at school I was not going to make it to graduation with a single unbroken tooth. He said that he thought a summer spent as a deck hand on a Mississippi River towboat might wean me from what he considered to be my mother's overly protective influence and toughen me up a bit.

"You'll meet a whole different kind of people than you've been meeting in Sunday School," he promised.

Two days later at midnight I boarded the "Lady Rhee" in East Saint Louis, Illinois, to start a three month tour. Our boat was too small to compete with the huge boats of the lower Mississippi and it was assigned to run back and forth between St. Paul, Minnesota and St. Louis.

> *In order for a son to want to stay on the farm or ranch, the father must be accepted as a success worthy of emulation.*
> *Stan Parsons*

The boat only stopped every three months to re-crew. Fuel and groceries were all taken aboard while the boat was in motion. Once you were on, you were on!

Upon my arrival I was told to go the boat's galley and await orders. In those days, the boats were run as quasi-military organizations.

Eventually, the 300 pound first mate who served as Master Sargent on the boat came in and sat down beside me.

He looked me up and down with disgust clearly registering on his face. "Kid, you know how to knife fight?" he asked.

When I answered in the negative he said, "Well, I'm here to teach you."

And he did. (With a butter knife.)

> *Deciding to have a business so you will have more time is like deciding to have a baby to have more time. The primary responsibility of any parent is to spend 100% of their time keeping the baby alive, happy and healthy. At some point babysitters, grand-parents and nannies can enter the picture, but at the beginning neither the parent nor the entrepreneur will have more time.*
>
> *Keith Cunningham*
> *Keystothevault.com*

As a young and tender man, he knew I was going to quickly have to learn to protect myself to survive on a boat whose whole deck crew (with my sole exception) had just been recruited from the county jail.

The pay was $12 a day for 12 hours of work plus all you could eat. No Saturdays or Sundays off. Most river town lawmen figured a criminal was as locked up on a towboat as he was in jail and they willingly supplied crews for a small "fee."

The man whom I shared a room with (opposite shifts) was a serial rapist who had become an expert in escaping from the Mississippi State Penitentiary.

"You put pepper in your socks so the dogs can't follow you," he told me.

Our cook was a Brit who had killed a man on a Cunard liner and was hiding out on the Mississippi. The boat owners loved him because he never wanted to get off.

"It was a fair fight. How can a fair fight be considered murder?" he asked me.

The deckhand I shared a shift with had a new, barely-heeled, pink scar that ran from the point of his shoulder down across his cheek and stomach to his hip bone from losing a knife fight on the dock at East St. Louis on his previous trip. He said he had to hold his intestines in his hands on the way to the hospital.

My father was right. They were different from the people in my Sunday School class.

Needless to say, three months later when I got off the boat I had developed enough self-confidence that I could "walk the walk." I was a man not to be messed with and it showed.

I never had another fist fight in my life.

(During the Mississippi River flood of 1973 the "Lady Rhee" hit the highway bridge at Greenville, Mississippi, and sank with the loss of all hands including my knife fighting mentor.)

Life and work with kids

When people ask what we do, we say we raise grassfed beef.

That's a bit misleading. What we really do is raise children, six of them.

On the ranch, we can both work at home with our kids, giving us ample opportunity for interaction, teaching and simply enjoying each other while doing work that has real and immediate rewards and penalties.

A farm or ranch is full of adventures and potential projects for kids, teaching them ingenuity, self-discipline and the satisfaction of a finished job.

Finally, a farm or ranch is a relatively inexpensive place to raise a family, with a reduced need for expensive stuff and diversionary recreational activities, and the savings that result from growing much of our own food.

Work is life and life is work. And, in the demands of work and life we can lose sight of the primary objective, which is to use this choice of life/work as a means to best raise our kids.

Caryl Elzinga, Idaho rancher

Sell your ranch to your children

In the business of borrowing and lending, the lender has the far easier job. Owner financing makes sense in that it spreads the return of capital over a long period of time and earns a lender's rate of interest rather than a saver's. That's about double the return in the USA.

Like the New Zealand sharemilking system, you should start out by gradually selling your child the livestock and let him build confidence in his income-earning ability.

I think it would be cruel to sell your ranch to your child at today's unrealistically high values. If your child wants to be a rancher rather than a real estate developer, you may want to get an agricultural easement to lower the price to its true agricultural value.

What most people don't understand is that an agricultural or conservation easement can allow you to cash in on the "bubble" of real estate prices as effectively as a sale by writing down the inflated value of the land over a period of years on your taxes.

Actually, by withholding some prime home-building sites that overlook your "long view" you can frequently have your cake and eat it too. While this trend is still in its infancy, land adjacent to large pastoral easements have shown huge increases in value.

Another option would be to sell your ranch to your child for 10 times its current earnings rather than at its appraised value.

I personally think that doling an inheritance out piecemeal while you are still alive is a far better option for lots of reasons. One of which is that being worth more to your children dead than alive has not proven to be very healthy.

Can a cattle ranch be a profitable, multi-generational investment vehicle? Gordon Hazard believes it can.

Hazard is who is well-known for his "tight but not cheap" approach to ranching in West Point, Mississippi, has a plan that will allow his ranch to continue to operate as a family investment for at least two more generations.

Currently, the Hazard ranch consists of 3000 acres of mostly fescue pasture in the Black Belt Prairie region of Eastern Mississippi. The Black Belt Prairie region features a shallow, black clay soil underlaid with limestone. This limestone keeps the land grassy and open despite the region's 60 inches of annual rainfall. It is premier cattle country.

The 3000 acres are divided into 40 paddocks for rotational grazing and support a herd of 1800 stocker cattle. The cattle are bought as lightweight calves primarily in the fall of the year and are grazed until the following August when they are sold as heavy feeders.

With very few cattle on the place after August, the fescue can be stockpiled for winter grazing and no hay is needed to overwinter the 1800 calves. This natural "unfair advantage" of cheap wintering combined with an average weight gain of 300 to 400 pounds per head has allowed the Hazard family to average a return over all direct costs of $100 to $130 a head for many years.

Gordon Hazard started the family ranch by buying 80 acres of land for $20 an acre while serving as an infantry officer during World War II. Offering to pledge his whole Army pay check as repayment, a local banker turned him down warning that buying a ranch "would be a millstone around his neck for the rest of his life." The banker across the street had a little sunnier disposition and lent him the $1600 needed to buy his first land.

This is the only loan the ranch has ever had.

After the war, he supplemented his ranch earnings by custom combining local crops with a pull type combine with a six-foot header. One day after his combine choked on vines three times in a row, he sat down and thought of jobs he could do that would involve cattle. He considered becoming an auctioneer but decided to become a vet instead.

Using the GI Bill he went to veterinary school at Auburn University, and worked for a short time for the State Veterinary Office. Recalled to the Army during the Korean War he became a Major in the Army Veterinary Corps.

> *You should develop a definite career path for your successor with both formal and informal training. Informal training comes from letting them have enough rope to make their own mistakes. The only way to teach your children ranch management and entrepreneurship is to give them their head and let them run with a ball.*
>
> *Stan Parsons*

Returning to Mississippi after the war, Gordon built a lucrative large-animal practice that eventually became one of the largest in the Deep South. However, despite his success as a veterinarian he still really wanted to be a cattle rancher.

Starting with 35 cows and a bull, he re-invested the ranch profits each year in more land and cattle. Realizing that stockers provided a higher return to the land he soon switched to stocker cattle. Using leased land to free cash flow for cattle purchases, he grew entirely from retained profits.

Using tax-free land exchanges, he was able to swap his original properties near town for much larger rural acreages, which speeded up the process of assembling a large ranch.

In those early wealth building years, he had to check his own cattle by the lights of his truck after a full day of working on other people's cattle. He said it was tough working two jobs but he never lost sight of his goal of assembling a large cattle ranch.

"I promised Sara that the day I had a thousand steers and the land it took to graze them paid for I was quitting veterinary practice," he said.

That day arrived in 1978 when Gordon was 55.

He was totally burned out as a vet and was more than willing to sign a non-compete clause with the vet who bought his large practice. Thanks to the payout on his veterinary practice and occasional consulting income, he never needed any of the ranch's income for living expenses and was able to reinvest all of it in additional land and cattle.

In 1978, he formed a C corporation to own the ranch's stocker cattle. The Hazard Cattle Company's five stockholders consisted of Gordon and his wife and their four children.

Gordon said he felt that C corps are best for stocker operations because there is no tax sheltering depreciation as there is with cow-calf. He said that C corporations are particularly attractive for cattle companies such as his where no funds are being taken out as a salary and the plan is for all of the profits to be retained within the corporation.

While cattle ownership definitely needs the liability protection of incorporation, land is usually best held by an individual. The cattle company then pays out its profits as a grazing lease payment to the landowner. Since rent is not subject to Social Security and FICA taxes, this arrangement at one time offered significant tax relief by avoiding the payment of Social Security and FICA taxes. Today, an "arm's length" separation must be proven between the landowner and the owner of the cattle for this tax dodge to be legal.

> *Ideas often lie unused because people do not want to use them. The fact that something is possible does not mean that it will happen.*
>
> *Thomas L. Friedman*
> *The Lexus and the Olive Tree*

This arm's length separation only applies to farming and ranching. Doctors and lawyers are able to pay themselves rent on their offices as always. This is highly discriminatory.

With inflation soaring in the late 1970s, Gordon and Sara realized that their ranch land presented a significant inheritance tax consequence to their children. Starting in 1979, they began to gift their land holdings to their children on a piecemeal basis at the maximum rate allowed by the law.

The current law allows you to give $11,000 a year tax free to any individual. This means that a husband and wife can give each of their children (and grandchildren) $22,000 worth of land or assets a year tax-free. Land and business assets must be gifted at their present-day value. Estates are now tax-free up to one million dollars.

Gordon said that because his children were still young he initially made them sign a 20 year lease to Hazard Cattle Company that would prevent them from selling their property. Later, confident that his children had inherited his frugality, he no longer required this.

"I wanted to give my children their inheritance while I was still living so I could have the fun of seeing my children do good and succeed," he said.

Today, all of the ranch assets have been transferred to his children and he currently serves as the ranch manager at no pay. The ranch has one part-time, occasional cowboy.

> *The father and son Aussie split*
>
> *Split the farm's land 50/50 with your son.*
>
> *Split the livestock business three ways between the father, the son and the son's wife.*
>
> *The livestock business pays going rental rates to the land business.*
>
> *The land business pays overheads and improvement expenses while the livestock business pays for direct expenses associated with livestock.*
>
> *Australian Farm Journal*

He said an 1800 head stocker operation is a nice size for a part-time grazing operation. He goes to the ranch at 7 AM every day and is through in three hours. On days when cattle are being wormed or shipped, his work day might extend to 11 AM.

He said the ranch consistently earns between $60 and $80 an acre for reinvestment. Due to what Gordon currently considers is a too high price for land in his area, these profits are now being invested in financial instruments rather than to expand the ranch.

He said that land in the Mississippi prairie had a history of being highly erratic in its value. Ranch land sold for $1000 an acre in the early 1980s and $175 an acre a few years later.

He said the ranch cost an average of $240 an acre to acquire.

Based upon its original cost, the ranch had returned 25 to 30 percent a year to the family. This means the family's net worth has doubled every two and one-half to three years.

Gordon said, "Part of your planning as a rancher is to have a plan to move your ranch down through the generations in an orderly and financially sound way."

> *The biggest thing most of need is accountability. We need someone to hold us to the things we say are important to us.*
>
> *Chris Le Pak*

To that end Gordon has also been training a replacement for himself in his grandson.

Mark Jr. studied at nearby Mississippi State while running a herd of 70 head of stockers of his own.

Mark Jr. started working on the family ranch as summer help when he was nine. Gordon said he gave Mark Jr. the option of taking his first summer's pay in cash or in the form of a steer.

"I had planned on paying him a hundred dollars but it wound up costing me $500 when little Mark opted for the steer," Gordon said.

When the steer reached 800 pounds Mark Jr. figured out that he could buy back two calves for the yearling and he did. And he kept doing so.

"When he got up to 30 head I told him he needed to be grazing them on his Daddy's ranch and not mine," he said.

Mark Sr. bought 160 acres of former CRP land for his son, and planted the ranch in fungus-free fescue and Max Q fescue.

Gordon hasn't set a date for his formal retirement from managing the ranch.

"I'm still making plans, establishing pastures and building new fences. I've loved cattle all my life and have never wanted to do anything else."

However, for all the romance, Gordon has told his children that a ranch is first and foremost a business and it has to earn a profit.

"I've told my children that if the ranch ever turns unprofitable, they should sell it. I told them don't keep it as a memorial to me if it's not making money."

Making room for your children

Dan Wilson had farmed conventionally for 28 years raising corn, soybeans, oats, barley, flax, alfalfa and hogs but decided it was time to switch to ruminant-based farm. In a family meeting, the Wilsons decided this transition would be an orderly one which would entail first becoming organically certified and then gradually switching to grazing animals.

Dan had promised his oldest son, Torray, that if he came back to the farm to live after college he would convert the whole farm to Certified Organic production and they started doing so with 240 acres in 2006.

Torray had added a small sheep flock while he was in high school on ten acres of flood-prone land. After returning from college in 2005, the sheep pasture was expanded to 75 acres.

The Wilsons were early converts to pastured sows and were the fourth farmer to sign up with Niman Ranch Meats in 1998.

"A neighbor asked me why I run my sows on pasture, I told him it was because I wanted to be able to breathe," Dan said.

The Wilson farm also has pastured poultry, both eggs and broilers, and bees.

"Our neighbors can't wait to see how much more weirder we are this year than last year," Dan Wilson said with a laugh.

All of the farm's lambs, eggs, chicken, honey and some pork are sold through the Shelburne Farmers' Market. Lorna,

Dan's wife, grinds her own wheat and makes bakery products to sell as well.

"There's just so much more opportunity with direct marketed foods than there is in commodity farming," Lorna said.

"Dan and I have five children and we'd like all of them to be able to come back to the farm.

"We think with our current direction, they can."

Lorna said she had always been interested in the concept of food as medicine and so has been very encouraging of the farm's conversion to pasture based enterprises.

"I see a pretty landscape as one of our most important products," she said.

"When you seed down 80 acres of corn ground to pasture, it makes a big difference in how pretty the farm looks."

Reflecting the high fertility and productivity of the farm's soil, the current plan is to manage it in an Argentine style pasture/crop rotation. The land will be in leguminous permanent pasture for five to seven years and then plowed down and planted to organic corn, soybeans, and a small grain nurse crop to a return to permanent pasture. They are currently converting 80 acres to permanent pasture each year.

The 2006 drought sped up the Wilson family's planned conversion. This came about when Torray attended a Holistic Management meeting at Holstein, Iowa, custom grassfed beef finisher Tom German's house in mid-July.

German told Wilson that due to the drought he needed a place to send some of his steers.

"After about five minutes of discussing what could be done, I realized we had the solution at our place.

"Now 49 head of grassfed steers are gaining fabulously grazing on some of our weedy, organically raised corn."

The second day the steers were on the farm, a neighbor came to the farm house in an absolute panic because he thought his cattle had gotten into the Wilson's corn.

"Dad asked me if I knew CPR because I might need it

to resuscitate our neighbors."

Dan said the farm is completely self contained in that all the feed used on the farm is grown on the farm.

"The only things we buy are vitamins, minerals and soybean meal."

Also, all of the soil nitrogen needed for the crops is produced from plowed down legume rich hog pasture.

"Our hog pastures are a mixture of alfalfa and orchard with Ladino clover," Dan said.

"But, there's only so much you can do with hogs on pasture.

"We want to put in ruminants as our centerpiece enterprise."

Torray said his dream is to have a pasture based, seasonal dairy as this centerpiece.

"The thing I really like about grazing is that one guy can do so much," Dan said.

"We now invest in animals instead of equipment."

A rancher's son creates a ranch of his own

David Evans grew up on a cattle ranch on the beautiful Point Reyes peninsula which juts into the cold Pacific Ocean just north of San Francisco.

After he graduated in 1996 from Cal Poly, he returned home full of vim and vigor and ready to turn the family ranch into a showplace example of management-intensive grazing. However, after two years of working for his parents he realized that they were happy with the traditional way of ranching and really didn't want to change things.

In 1999 he decided to start something of his own and leased 110 acres and bought 80 yearlings. These stocker cattle produced fantastic gains of 2.5 lbs. per day but he wound up losing $1000 due to negative marketing margin.

"I learned the pitfalls of the commodity business real quick. I decided I wanted to get into something where I could call the price of what I produced," he said.

The next year he sorted off two heifers from his stocker herd as a test of grassfed beef.

"I thought the beef was great, but then I was pretty biased."

In 2001, he thought he would be able to direct market five head locally. However in a classic example of timing and luck, he had purchased a listing on Eatwild.com two days before Michael Pollan's article "Power Steer" appeared in *The New York Times Magazine.*

"I got calls all the way from New Jersey and wound up selling 25 head rather than five head," he said.

Evans said his dream was to get his beef into world famous Chez Panisse Restaurant in Berkeley.

"You get in there and you have instant credibility for your meat's quality."

Unfortunately, his meat got a less than enthusiastic reception. He said the chefs at Chez Panisse gave him excellent feedback on the quality of his beef and what was lacking.

"Like most producers I started out doing it all wrong. I got in a hurry and was harvesting my beeves before they were really finished.

> *The reason it is so hard to follow the leader is that the leader got there by doing something remarkable. And that remarkable thing is now taken and is no longer remarkable when you do it.*
>
> *Seth Godin*
> *The Purple Cow*

"Once I started harvesting cattle at 24 months of age all the restaurant doors suddenly opened for me. Chez Panisse, Acme Chophouse. I got them all."

Today, white table cloth restaurants are 75% of his sales but that is getting ahead of the story.

As a part of his emphasis on meat quality, Evans wanted to cut his own meat. Unfortunately, zoning ordinances in Point Reyes do not allow new use butcher shops.

"The only way I could have a butcher shop was if it was in conjunction with a restaurant. So, I had to open a restaurant."

The small 25 seat restaurant is called "Marin Sun Farms Eatery."

"We were primarily a retail meat shop, but for $10 we would cook your steak and serve it with locally grown organic vegetables."

He said that the restaurant was a marginal enterprise but that it was worth it to get the butcher shop.

"Cutting your own meat is the only way you can control its quality. My customers are very particular about how a piece of meat looks on the plate."

With a new 1700 acre leased ranch, all the pieces of the puzzle were finally together and Evans put the pedal to the metal and started growing as fast as possible.

"I was enjoying the first out of the gate advantage. I quickly established my brand with the top restaurants and virtually no competition."

He hired 17 employees. He rented a booth at the Ferry Plaza Farmers' Market, started selling 600 to 800 pounds of ground beef a week to Stanford University and most of his offal went to two pet food companies. In the year 2005, he harvested 265 head compared to only 110 the year before and 60 the year before that. He said he felt he was on track to harvest well over 300 in 2006.

And then disaster struck.

On New Year's Eve 2005, the small river adjacent to his butcher shop flooded.

"I thought it was the end of the world, but in retrospect, it probably saved my life.

"I was growing entirely too fast. I didn't have time to think."

Luckily, Evans had flood insurance which largely made him whole financially but the flood has resulted in a dramatic downsizing.

> *If you are out of time and out of money, you can't afford the tuition to learn something new.*
>
> Gregg Simonds

Evans now has only four employees and the restaurant has been rented to a local catering company. He said his current goal is to probably cap his growth at 10 head a week.

He said he has figured out that growing slowly is actually the only way to be profitable.

"You need to take the time to build a good foundation for a business.

"You get one or two head a week ahead of your sales

and you've suddenly got a problem that can be really tough to get over.

"We'll be down slightly in gross this year but we will actually make much more money."

When the thrill is gone, find a new challenge

What is a common emotional reaction in men to having achieved success?

According to Dr. Stephen Berglas of Harvard Medical School's Department of Psychiatry, the answer is frequently acute depression.

This is because for most entrepreneurs "getting there is all the fun." Berglas wrote an excellent article on this in *INC Magazine*.

He said that what keeps us going in the early struggle years of developing a business is a desire to show all those people who think we are crazy that we aren't.

"During the start-up phase, you're essentially working to gain self-esteem," he said.

"You're working for bragging rights. Daring the devil and beating him makes you feel smart, confident and powerful. Eventually you'll reach a point of optimal competency."

This is often called the "sweet spot." This is when your challenges and competency are in perfect balance. However, Berglas said that this is a very short moment in time and that it never lasts.

"After you've attained a measure of success, you typically begin shepherding the business rather than creating, developing or expanding it.

"And there's the rub, because a sense of creating is the core psychological gratification that you as an entrepreneur are after. When you've obtained the pinnacle of success you've been striving for, your sense of challenge, creativity and drive will cease to exist."

Once this drive ends, he said you are at the point of your greatest psychological risk of depression, often described as burnout and burnout is very different from stress.

He said stress comes from demands that are beyond your sense of competence.

Burnout comes from facing goals that are too readily achieved or have already been met. He said the only way to avoid burnout was to continually create new challenges to overcome.

"Unless you figure out how to put challenges into your life on a regular basis, burnout will follow."

And denying the existence of burnout can lead to some life-wrecking decisions. He calls these the "The Four A's." These are aloneness, arrogance, adventure seeking and adultery.

"If you're a successful entrepreneur, you're usually alone at the top. But when you start feeling arrogant, negative consequences are inevitable.

"Rather than seeking challenges or opportunities to dare the devil, you seek

Six tips for happiness
1. Give yourself permission to be human - emotions are natural.
2. Engage in activities that are both personally significant and enjoyable.
3. Happiness is dependent upon what we focus on. Do you see the glass as half full or half empty?
4. Simplify your life. We compromise our happiness by trying to do too much.
5. What we do with our bodies influences our mind. Healthy eating leads to both good physical and mental health.
6. Appreciate and savor the good things in life from your food to your friends.

Tal Ben-Shahar
Harvard University

adventure to spit in the devil's face. That in turn, blinds you to the need for people as supports. With an arrogance-fueled sense of self, committing adultery is a relatively easy thing to do."

Berglas said what you are missing is that feeling of competency feedback that you got in the early stages of your business' development. You were getting better every day and it showed.

In ranching, this challenge is perhaps best met by adding a new species of livestock or an enterprise that taps into and exploits the unmet psychological needs of the urban population.

Before you think your ranch is totally maxed out, take a look at the enterprise list Joel Salatin came up with for his operation. Try making a similar one for your place.

The secret is to try to build on what you have already built in terms of personal skills. What you are seeking is eustress not stress.

Eustress is defined as seeing just how high you can raise the proverbial bar. Again, what you are seeking is proof that you are getting better.

Eventually, Berglas said you will fully exploit all of your options for diversification within your existing business but you can't quit seeking challenges. This is the entrepreneur's curse.

An entrepreneur absolutely must find new challenges to maintain good mental health.

Teaching others what you have learned can be as stimulating as learning something new yourself.

"When you teach, you get feedback that reaffirms your competence and increases your self esteem," Berglas said.

He said this teaching can be to people inside or outside of your business. Both increase self-esteem.

Another challenge is to prepare to pass your business on to others by redesigning it so as to eliminate all those tasks you do not like to do. This can be by changing the business to eliminate those tasks or by delegating them to someone else.

Another good challenge is to seek out and develop relationships with other successful people. He said entrepreneurs need the intellectual stimulus of peer give and take.

"Networking with peers will ensure you'll get feedback from people who aren't intimidated by your stature."

> *Every problemm has numerous possible solutions. The key to success is never about what you do, but rather about how you do it.*
>
> *Most people don't have what they want in their lives because they consistently select strategies that they cannot execute. Too many people select a solution which is not consistent with nor aligned with the skills they have and thus they can't execute it and it fails.*
>
> *People who consistently achieve their goals and fix their problems look first to their skills and capabilities. The strategy you select should be based on the skills you have available or that you can quickly acquire and that you can actually execute.*
>
> *Keith Cunningham*
> *Keystothevault.com*

Finding a political cause outside of your business that you care deeply about can often reinvigorate your passion for life.

"Writing a check will not correct burnout. Instead, use your power, position and skill to fight for a cause you're passionately committed to."

However, the one thing Berglas said you should never do is to ignore feelings of burnout. "I promise you, the burnout you are feeling will only increase. Psychologically, you must have meaningful goals that give you an internal season of well-being and self-esteem.

"After working with scores of people who have earned more money that they dared dream of when they were starting their companies, I'm positive that those feelings cannot be bought at any price."

Ranching after 50

I once read that what makes it so hard to foresee the future is that we can't foresee ourselves changing.

In other words, we believe that the way we are is the way we always will be. That we do change can only be seen in retrospect.

Every day I look in the mirror and increasingly see my father who died at 64.

Every day I can relate more and more to that old 1960s saw that said, "If I had known I was going to live this long I would have taken better care of myself when I was young."

However, everything that happens to us after 50 can't be avoided by a healthy lifestyle and exercise. It is just something you have to go through to get to the other side.

I recently tried to tell a 35-year-old man what would happen to him when he turned 50 and he just grinned.

He jogged five miles every day and planned to stay in shape. He wasn't going to have any mid-life slump he insisted.

That one's midlife changes are caused by hormones with their own pre-set clock, is something that no young man can believe. When it happens it is always a surprise and then it gets scary.

One of the things the male hormone does is to allow you to handle stress. A young man in combat can still think clearly. This is why soldiers are drafted at age 18 at the peak of male hormone production.

As your male hormone level declines with age, you either lower the level of stress in your life to match your testosterone production or you die of a heart attack. A doctor told me once that the years 50 to 56 are the most deadly in a man's life for first-time heart attacks.

> *The source of continuing aliveness is to find your passion and pursue it with your whole heart and mind.*
>
> *Gail Sheehy*
> *Passages in Men's Lives*

The problem is that most of us arrive at 50 with our stress wagon loaded to the breaking point trying to maximize income. The absolute necessity to dump this load as quickly as possible before it kills us scares the heck out of our loved ones.

To them we have gone "Middle-aged crazy."

This sudden desire for escape also scares the heck out of employers. Very few white-collar, upper-income men make it to retirement age today working for the same company they worked for in their 40s.

Most get "retired" in their early 50s when they can no longer maintain the pace their income level requires of them to be profitable to their employers.

Self-employed men whose income is directly tied to their personal effort typically see their income peak around age 53. As a result, on average, most employed and self-employed men's income at age 60 is approximately the same as it is for men aged 30.

Such an upside down U-shaped income curve is hardly what most men plan for. My father's career followed the exact trajectory I have described. He was able to parlay a very narrow agronomic knowledge specialty into something of great value to the Department of Defense during the Vietnam War.

The ranch was turned over to my mother and me to run and Dad largely disappeared into a shadow world he couldn't discuss with anyone.

He lived in a world of international travel, tailored suits, a full-time personal assistant, a new company car every year, a six figure income plus an unlimited, no-questions-asked, expense account. However, the pace of the job he built for himself in his 40s almost killed him in his early 50s when he had two major heart attacks.

He was offered a lower stress job but insisted on the same income as payback for "past services rendered." When his employer refused, he became intransigent and got fired.

That this could happen to him came as a huge shock.

That he could not find another job at anywhere near the same money at age 55 was an even bigger shock.

That a man who had taught college at 27 could only find a job teaching part-time in a high school at 55 was inconceivable to him.

> *The ten wealthiest people on the planet are all still working. Not one of them is concerned with financial freedom or retirement because they love doing it.*
>
> *Keith Cunningham*
> *Keystothevault.com*

As financially successful as he appeared, in the end all he had was a great job.

To compound his misery, my mother died shortly after he lost his job. She had kept his home world of ranching going and had always covered his bets when his hobby of leveraged speculation required it. Now he was totally adrift.

Dad quickly sold the ranch for a small fortune, quit his teaching job and re-entered the dating world as a very prosperous looking, sophisticated man of the world.

Even when he was at the top, I think in his heart he knew it wouldn't last. Like a lot of fathers he tried to relive his life in his son.

He told me again and again, "You take care of business as a young man, so your business can take care of you as an old man."

He told me to find out what kind of business I wanted and he would lend me the money to get it started. Unfortunately, he died before I found out what that business was.

Because of a late-in-life marriage to a much younger woman, I never received an inheritance from him. When his second wife died I had to pay for her funeral. Dad's money had been long gone.

But, my father had successfully bent the twig. I knew the last thing I ever wanted in life was a good job.

Now, I would like to tell you that I instilled that same entrepreneurial spirit in my children but apparently I haven't. Both of my daughters work for the government — one as a hospital pharmacist for the Veteran's Administration and the other as a school teacher.

I read that entrepreneurship is a "generation-skip" kind of a thing. In other words, the children of entrepreneurs typically aren't entrepreneurial but their grandchildren are.

I have six of them so far, but only time will tell if the old saw holds true.

The early 50s hit me really hard but I feel better and better every day. For those of you who are just entering the 50s, I advise you to hold tightly to the people and dreams that are most dear to you.

Do what you have to do to downsize the stress in your life but do it without chunking everything and everybody overboard in the storm.

Try to explain to your wife and children that you aren't crazy, you're just middle-aged and that going a little crazy is a part of being a 50-something-year-old man.

Hold and roll. You will come out the other side. And, while you can't foresee it now, you will be different. Hopefully better, certainly wiser.

If we put as much time and energy into creating a life worth living as we do in pursuing the seemingly holy grail of financial freedom, we would be stunned by the results we would get in our lives.

Financial freedom is an outcome — an effect! The key to having what you want in life is to fall in love with the causes. Why not fall in love with what you are doing and give yourself to it 100% and see what it gives back to you.

This lifetime is not solely about getting to financial freedom, but rather to live it so that at the end you can say, "This was my life. I am proud of it and I would gladly live it again if given the chance."

Keith Cunningham
keystothevault.com

Bibliography

Cunningham, Keith. *Keys To the Vault, Lessons from the pros on raising money and igniting your business.* Enclave Publishing, 2006.

Dolan, Paul. *True to Our Roots, Fermenting a Business Revolution.* Bloomberg Press, 2003.

Drucker, Peter. *The Daily Drucker.* Harper Business, 2004.

Friedman, Thomas L. *The Lexus and the Olive Tree, Understanding globalization.* Farrar, Straus, Giroux, 2000.

Gerrish, Jim. *Management-intensive Grazing, The Grassroots of Grass Farming.* Green Park Press, 2004.

Godin, Seth. *The Purple Cow.* Portfolio, The Penguin Group, 2002.

Judy, Greg. *No Risk Ranching, Custom Grazing on Leased Land.* Green Park Press, 2002.

Kiyosaki, Richard. *Rich Dad, Poor Dad, What the Rich Teach Their Kids About Money That the Poor and Middle Class Don't.* Teachpress Inc., 1999.

Kiyosaki, Richard. *Prophecy, Why the Biggest Stock Market Crash in History is Still Coming...and How You Can Prepare Yourself and Profit from It.* Warner Business Books, 2002.

Mahar, Maggie. *Bull! A History of the Boom 1982-1999.* Harper Business, 2003.

Nation, Allan. *Knowledge Rich Ranching.* Green Park Press, 2000.

O'Hara, William T. *Centuries of Success.* Adams Media, 2004.

Peters, Tom. *Liberation Management, Necessary Disorganization for the Nanosecond Nineties.* New York: Ballantine Books, 1996.

Rapaille, Clotaire. *The Culture Code.* Broadway Books, 2006.

Rogers, Everett M. *Diffusion of Innovation.* New York: The Free Press, 1962.

Salatin, Joel. *Family Friendly Farming, A Multi-generational Home-based Business Testament.* Polyface, Inc., 2001.

Senge, Peter. *The Fifth Discipline, The Art and Practice of the Learning Organization.* New York: Doubleday, 1990.

Sheehy, Gail. *Passages in Men's Lives: New Directions for Men at Midlife.* Simon & Schuster, 1999.

Shilling, Gary. *Delfation, How to survive and thrive in the coming wave of Deflation.* McGraw-Hill, 1999.

Slyotsky, Adrian. *The Art of Profitability.* Warner Business Books, 2002.

Stewart, Thomas A. *The Wealth of Knowledge, Intellectual Capital and the Twenty-First-Century Organization.* Currency, Doubleday, 2001.

Stockman Grass Farmer Magazine.

Ward, John. *Family Business Review Magazine.*

Yegge, Wilbur. *A Basic Guide for Valuing a Company.* John Wiley & Sons, 1996.

Index

About the author

Allan Nation has been the editor of the *Stockman Grass Farmer* since 1977. The son of a commercial cattle rancher, Nation has traveled the world studying and photographing grassland farming systems. He received the 1993 Agricultural Conservation Award from American Farmland Trust for spear-heading the drive behind the grass farming revolution in the USA. He has served as a consultant and resource for Audubon Society Television Specials, WTBS, PBS, and National Public Radio. He is also the author of *Pa$ture Profit$ with Stocker Cattle, Quality Pasture, Knowledge Rich Ranching, Grassfed to Finish* and other books and publications. He and his wife, Carolyn, live in rural Mississipppi.

Name ____

Address ____

City ____

State/Province ____________________Zip/Postal Code_____

Quantity	Title	Price Each	Sub Total
____	**Farm Fresh** (weight 1 1/2 lb)	$30.00	________
____	**Grassfed to Finish** (weight 1 1/2 lb)	$33.00	________
____	**Knowledge Rich Ranching** (weight 1 1/2 lb)	$32.00	________
____	**Land, Livestock & Life** (weight 1 1/2 lb)	$25.00	________
____	**No Risk Ranching** (weight 1 1/2 lb)	$28.00	________
____	**Management-intensive Grazing** (weight 1 1/2lb)	$31.00	________
____	**Pa$ture Profit$ with Stocker Cattle** (1 lb)	$24.95	________
____	**Quality Pasture** (weight 1 1/2lb)	$32.50	________
____	**Shelter & Shade** (weight 1 lb)	$20.00	________
____	Free Sample copy ***Stockman Grass Farmer*** magazine		

Sub Total ________

Postage & handling ________

Mississippi Residents add 7% Sales Tax________

U.S. Funds Only, Please TOTAL ________

Shipping:	Amount	Canada
Under 2 lbs	$5.60	$8.50
2-3 lbs	$7.00	$9.60
3-4 lbs	$8.00	$11.25
4-5 lbs	$9.60	
5-6 lbs	$11.50	
6-8 lbs	$15.25	
8-10 lbs	$18.50	

Foreign Postage: Add 40% of order.

We ship 4 lbs per package maximum outside USA.

Green Park Press books and the ***Stockman Grass Farmer*** magazine are devoted solely to the art and science of turning pastureland into profits through the use of animals as nature's harvesters. To order a free sample copy of the magazine or to purchase other Green Park Press titles:

Please make checks payable to

Stockman Grass Farmer
PO Box 2300
Ridgeland, MS 39158-2300

1-800-748-9808
or 601-853-1861
FAX 601-853-8087

THE STOCKMAN GRASS FARMER

GREEN PARK PRESS